The Bombed Buildings of Britain

a record of architectural casualties: 1940-41

edited by J. M. RICHARDS
with notes by John Summerson

Contents

FOREWORD	2
LONDON	5
BRISTOL (AND CLIFTON)	81
COVENTRY	91
PORTSMOUTH	97
PLYMOUTH	101
MANCHESTER	107
LIVERPOOL	111
OTHER LARGE TOWNS	117
COUNTRY (AND SMALLER TOWNS)	127
INDEX	139
ACKNOWLEDGEMENTS	140

Foreword

THIS book illustrates the damage that was done to buildings of architectural note during the heavy raiding of this country from the air between August 1940 and May 1941, and the spasmodic raiding during the rest of 1941. It does not claim to be an exhaustive survey of all the architectural damage done. Perhaps after the war such a survey will be undertaken officially and comprehensively, though I am not sure that it would have much meaning ; for in the perspective of history we see buildings rise and fall for a multitude of reasons, and it is only of passing interest to single out one of them as in a class by itself. Storm and lightning, the death-watch beetle, Cromwell's troopers, the speculative builder, mere obsolescence—and now German bombs ; the legacy they leave of ruins, or living architecture reduced to memories and legends, is all one.

But this passing interest is an intense and legitimate one. After the smoke and dust of the bombing has at least temporarily cleared away, it is natural that we should want to know what of real architectural value has in fact been lost, and to have the history and character of these buildings recorded so that they shall not have vanished without some kind of obituary tribute. Besides this there is the pictorial aspect of bomb damage, the interest we take for their own sake in the ruins that suddenly changed so many familiar skylines overnight. These ruins have already lost much of their vivid spectacular character in the tidying-up process, and will eventually become only a memory. Photographs of them are part of the documentation of contemporary history.

This book, therefore, has two purposes : to provide an obituary notice and a pictorial record. To appreciate the latter of these, one must consider each ruin as an architectural phenomenon in its own right. At first mention it may appear unfeeling that the connoisseur of ruins should regard as material for objective appraisal scenes which for most of us symbolise the horror of lives lost and irreplaceable treasures destroyed. But it is surely the very intensity of the symbolism they are charged with that justifies the claim of ruined buildings to be looked at for the sake of what they are at the moment. It has always been the role of the ruin to compress into the same picture both the embodiment of historical experience and the form and colour of architecture itself. Hence its romantic appeal.

The public mind, without necessarily deluding itself that air-raid destruction has of itself done much to bring improvement nearer, has universally identified the destruction of the congested centres of our cities with the possibilities of reconstruction. The destruction has been made to symbolise the spectacular end of an era that will not return. From this point of view the ruin, even when looked at as architecture in its own right, represents the apotheosis of the past—the intense experience of these active days crystallised into architectural shape. This is not the place to discuss the question of which damaged monuments should be restored and which demolished. Each case must be decided on its own merits, topographical, social and architectural. But when it is all over a few of the bomb-wrecked buildings might well be left as permanent ruins—not, one hastens to add, as object-lessons for future war-mongers or for any other moral purpose—but for the sake of the intensely evocative atmosphere they possess in common with all ruins, which gives them an architectural vitality of their own ; and frankly for their beauty. To posterity they will as effectually represent the dissolution of our pre-war civilisation as Fountains Abbey does the dissolution of the monasteries.

Here, then, on the following pages, are some pictorial records of the ordeal architecture suffered in 1940-41. These, with our memories of that peculiar air-raid smell of wet charred wood, of the blundering gait with which we picked our way over puddled streets criss-crossed with hoses on dark winter mornings, and of the familiar houses we saw splintered with impressive thoroughness into a spillikins heap of dusty timbers, may together form a background of smells, sights and sounds sufficient to evoke for us the whole strange aftermath of bombing. This aftermath, it need hardly be emphasized, is of a different pictorial character from the bombing itself. This is not a picture-book of air-raid scenes. Instead of the confused, dynamic drama of active destruction, and the human heroism that went with it, here is its architectural by-product, the residue left high and dry after the wave of destruction has passed on. Its quality, by contrast, is altogether impersonal and static, even reposeful, despite the fact that it takes much of its distinctive character from the suddenness of destruction's onslaught, a very different affair from the imperceptible, natural process of decay which has produced the mouldering ruins the connoisseur of architecture already knows so well.

The architecture of destruction not only possesses an aesthetic peculiar to itself, it contrives its effects out of its own range of raw materials. Among the most familiar are the scarified surface of blasted walls, the chalky substance of calcined masonry, the surprising sagging contours of once rigid girders and the clear siena colouring of burnt-out brick buildings, their rugged cross-walls receding plane by plane, on sunny mornings in the City. Moreover, the aesthetic of destruction bears no relationship to any architectural merit the building may have possessed in life. In death merit is of a different order, and some of the most dramatic and evocative ruins have flowered suddenly out of a structure no one would have looked at twice.

These pictures, however, are limited to buildings that were of architectural importance in life, buildings whose loss is more profound than any transient beauty in their swan-song can compensate for. This brings us to the other of the two purposes I said this book is meant to serve, and really, of course, the chief one : to put on record those of our national architectural possessions that have become casualties, to illustrate the nature of the casualties and to accompany the illustrations with some notes about the character and history of the buildings concerned. In some cases a secondary purpose may be served when a photograph of a building in its damaged state happens to throw some new light on its design or construction, but the pictures have not been selected with this technical end in mind.

The criterion of architectural merit and historic interest is not itself susceptible of exact definition. In this sense the selection of buildings is a personal one, but the aim has been that it shall be as representative as possible and, though completeness is not claimed, I do not think, as a matter of fact, that many buildings of obvious architectural importance that have been badly damaged are missing, except for one or two that may not be mentioned for security reasons. Incidentally, because of the habit bombs have of falling alike on the just and the unjust, this collection provides a fair summary of the course of English architecture—at any rate so far as urban building is concerned.

As a final point in defining the scope of the book, it should perhaps be mentioned—though it is probably sufficiently obvious—that in order to keep the material within bounds, buildings are generally only included when damage has been severe. Broken windows, holed roofs, scorched walls and scarred stonework—common sights now in nearly every part of the country—are not taken to justify inclusion.

Anything like completeness—as distinct from representativeness—is of course impossible in the case of domestic architecture. A tragically large number of the decent, sound (generally anonymous) terraces, squares and individual houses that form so characteristic a part of the English architectural heritage, and give what unity they possess to most towns and cities, must have been destroyed. I am thinking, in particular, of houses of the eighteenth and early nineteenth centuries. It is impossible to record them all, or even all of outstanding architectural quality. So on the following pages a few fairly well-known examples must be taken to represent the mass of obscure examples of each period and type. Various parts of Bloomsbury, the Inns of Court, and Finsbury Square in London, Berkeley Square, Bristol, Albion Street,

FOREWORD

Hull, and one or two others represent the aristocratic domestic planning of the Georgian age. Park Crescent, London, Lockyer Street, Plymouth, and a Brighton terrace represent the stucco domestic architecture of the Regency, and Portsmouth and Newmarket High Streets the informal but delightful sequences of shops, pubs and bow-fronted houses that survive as the civilised backbone of many provincial towns that have otherwise long since lost their shapeliness. Similarly with industrial architecture: a great many of the noble, simple early nineteenth century warehouses and factories of our seaports and industrial towns must have suffered badly. These are represented by only a few examples, for in this case a full survey could not in any case be published for security reasons. The same ban applies even more forcibly to naval dockyards, the vigour and character of much of whose architecture has, indeed, never been well known.

The arrangement of the book needs little explanation. The first section is given to London, the six shorter sections that follow it to the six provincial cities that have suffered most severely, the next section collectively to other big cities where architectural losses have not been so great as to warrant their being given a separate section each, either through good fortune as to where bombs have fallen or because they had less of architectural value to lose (I emphasise this, lest the inhabitants of these cities should suspect any belittlement of their actual sufferings: I know how, after the event, the extent of damage becomes a matter of local pride), and the final section to damage outside the big cities, in other towns and where stray bombs have struck unluckily in the country. Within each section the buildings are grouped first according to type (churches, public buildings, domestic, etc.) and then as far as possible in order of date.

Acknowledgment should be made to THE ARCHITECTURAL REVIEW, for which this material was originally compiled, and I would like to record my own thanks to Mr. Summerson for his notes, to which the credit will be due if this book is thought to have any more permanent value than that of a compilation of bomb-damage pictures; also for much help in collecting the material, given by him and by the Ministry of Information, the National Buildings Record and others who are acknowledged on the last page.

J. M. R.

LONDON

Aldermanbury, in the City of London.

Guildhall

GUILDHALL is one of those buildings where sentiment has outstripped discrimination and whose architectural value is apt to be assumed rather than discerned. It was originally a fine, sumptuous hall, built between 1411 and 1433, but seriously damaged by the fire of 1666. Wren restored it and gave it a clerestory and a flat ceiling. In 1789 the younger Dance added the Gothic façade towards King Street. Then, in 1866-70, Sir Horace Jones, the architect of Smithfield Market and Marshall and Snelgrove's, recast the whole building. It is chiefly Sir Horace's work which has suffered under the raids. The best that can be said for it is that it was good Gothic for a man obviously more at home with Italian and hardly a genius at either. The mediæval fragments of the Hall survived the fire of 1940 as bravely as they did that of 1666. The large picture shows the elder Bacon's monument to Chatham, one of the series of sculptural tributes to National Heroes, reared incongruously against the ancient walls.

6

City Companies' Halls

BREWERS' HALL, in Addle Lane. was unquestionably the finest of the Halls rebuilt after the fire of 1666. It was built by Captain Caine in 1670-73, the "model" or general plan being provided, apparently, by two members of the Company's court. The exterior, facing on to a little private courtyard, was characteristic of Restoration London; the stairway seen in the illustration was a rebuilding by Cubitt, 1 59-60.

HABERDASHERS' HALL, Gresham Street, was rebuilt after the Great Fire, but burnt again in 1838, except for the Court Room and Drawing Room. The Court Room had a very good plaster ceiling and there was some interesting seventeenth-century carving.

BARBERS' HALL, Monkwell Street, was remarkable for the particularly fine Court Room of 1636, with which the name of Inigo Jones has been associated. It escaped the fire of 1666, but has now quite disappeared except for a few carved stones.

City Companies' Halls

BAKERS' HALL was burnt in the fire of 1666 and again in 1714. Rebuilt in 1719-22, it had a rather good panelled hall, Court Room and staircase. James Elmes repaired and restored it about 1825 and it was extensively redecorated quite recently. The entrance from Harp Lane, seen in the picture, is typical of the kind of substitute built in the 'sixties and 'seventies to replace the vigorous heraldic frontispieces which gave consequence to these approaches in the plain streets of the Georgian city. This example is by J. Clarke and C. J. Shoppee, 1882.

ATIONERS' HALL, off Ludgate
l, is memorable for the very re-
ed late eighteenth-century façade
 the Hall itself, designed by
 C. Mylne. This is visible from
tioners' Hall Court, and has not
n entirely destroyed. Behind
s façade is a substantially seven-
nth-century interior with a fine
inthian screen. To the north is
 badly damaged block containing
 Court Room, richly fitted up in
 Palladian period. This is shown
 the photograph on the facing
e, which includes a portion of
 plaster overmantel and the
ture of a former Lord Mayor of
ndon who was Master of the
npany. Beyond the Hall are the
re-houses, which form the west
e of a small courtyard, to the
th of which is St. Martin's,
dgate. Stationers' Hall is not
 seriously damaged for recon-
uction to be out of the question.

PARISH CLERKS' HALL, built 1669-72, was
something of a curiosity in that it occupied only
the upper part of a building, the lower part being
let as offices. The street front and much of the
interior had been built in modern times. It
contained a little seventeenth-century glass and
panelling, but nothing of great importance.
To-day nothing remains. The Parish Clerks' was
among the humblest of the City Companies and
its Hall was correspondingly modest.

RDLERS' HALL, in a court off
singhall Street, was a very plea-
t building with particularly good
ving in the Hall and Parlour.
 was built to the designs and
imates of two tradesmen, Work-
n and Lowe, in 1680-2. The
dwork was by a Mr. Phillips.
y overmantel in the Parlour
tained a painting by Richard
lson, and some of the architec-
e was of eighteenth-century date.
 richly Victorian drawing-room
 added by Woodthorpe in 1878.
 most serious loss, however, is
 screen and gallery in the Hall.

FISHMONGERS' HALL, to the
west of London Bridge, remains
an imposing mass, in spite of
heavy damage. Designed by
Henry Roberts in 1831, it is
very much what one would
expect from a loyal ex-assistant
of Smirke : thoughtful, if a little
dull, leaving a good Ionic order
to speak for itself. The interior
is very splendid, with sombre,
opulent masses of neo-Grec
detail. Gilbert Scott made all
the working-drawings for this
building while in Roberts's
office.

City Companies' Halls

MERCHANTS' TAYLORS' HALL, the biggest and one of the most interesting, occupied a site between Threadneedle Street and Cornhill, locked in by modern office blocks. Its buildings surrounded a garden, and consisted of a Gothic Hall, much modernized, a kitchen, Court Room, Parlour, and other apartments, some mediæval, some seventeenth and eighteenth century, some mid-Victorian. The best work was in the staircase and Gallery, where the woodwork had a Hawksmore-like accuracy of design. All this has been burnt, and is more to be regretted than the Hall, whose mediæval carcase largely remains and, with careful examination, should reveal worth-while evidence of the buildings' remoter history. The Gothic windows seen in the photograph were designed by Edward I'Anson in 1878-9 and the timber roof was also a modern one ; the Ionic screen of 1672 is a regrettable loss.

Lambeth Palace

LAMBETH PALACE LIBRARY, originally the Great Hall where the Archbishops dined, is a curious specimen of Gothic built out of its time. It was in progress in 1663, the year in which Archbishop Juxon died. Juxon, nearly 80 when he became Primate, was a link with the "Gothic Revival" of Laud's day and his distinctly Laudian taste in architecture is represented in this last flourish of hammer-beams, spandrils and pendents. It is clear that Juxon's carpenter, perhaps not too familiar with this kind of work, took Westminster Hall as his model, building on a smaller scale and adorning his structure with classical mouldings. The book-cases, designed by Edward Blore, were inserted when the Hall was converted to the purposes of a library.

St. Paul's Cathedral

ST. PAUL'S CATHEDRAL has been twice hit by high explosive bombs. The two incidents—one in the choir, the other in the north transept—are illustrated above. The reredos, by Bodley and Garner, 1888, an elaborate but muddle-headed design which interferes with the clarity of Wren's architecture, was practically undamaged. The bomb which fell in the north transept wrecked the marble portico to the *north* door, which formed part of the old organ screen and bore Wren's epitaph, the famous : *Si monumentum requiris, circumspice.* The right hand picture shows the north transept with the wreckage of the portico. The mosaics *in* the spandrils are from designs by Alfred Stevens (left) and G. F. Watts. On the facing page is a closer view of the north transept, showing where the debris brought down by a bomb caused the church floor to collapse into the crypt. The large monument against the pier, damaged by the explosion, is to Admiral Lord Rodney and is the work of Rossi. The standing figure against the end of the pier is Earl St. Vincent, by Baily. Among the debris can be seen remains of the portico to the north door. It is fortunate that neither of the direct hits which the cathedral has received have interfered with the stablility of the structure as a whole, having merely pierced the light coverings between main points of support.

12

Churches

CHELSEA OLD CHURCH was that rare thing, a church which no restorer had ever spoilt. Century by century it had collected beauties of craftsmanship and design, and as it stood before the war it was perhaps the most human and eloquent of London's historic buildings. Parts of the church went back to the Middle Ages, but the brick body and the tower were built in 1667-70. Both were admirably designed, though in a surprisingly vernacular fashion, considering the date and the proximity to London. At the south-east corner was (and still is, in a shattered condition) the chapel built by Sir Thomas More, and containing what was to have been his tomb had he died a natural death. Of the many lovely monuments, it is satisfactory to know that all of great importance have been rescued from the ruins, many practically undamaged. The photograph shows the wreckage of the church from the south-east, with Sir Hans Sloane's monument in the foreground.

THE DUTCH CHURCH, AUSTIN FRIARS, has been demolished with spectacular thoroughness, only the eastern arches built into a party wall having survived. The building was the nave of an Augustinian church, handed over to the foreign Protestant community by Edward VI and subsequently assigned exclusively to the Dutch. It survived the Great Fire, but was partly burned in 1862 and hideously restored. The arcades and traceried windows were elegant specimens of 14th century work and are to be regretted along with many beautiful monuments of the 17th and 18th centuries. Under the débris is still a rich paving of mediæval and renaissance gravestones.

ALL HALLOWS, BARKING, was one of the nine City churches in which pre-reformation Gothic was represented. Its nave arcades were Early English, though much altered in the fifteenth century and later. The east window was an especially complete and happy example of the fourteenth century. Its delicate tracery has gone. The fifteenth century added the chancel arcades, the clerestory and the chapels. The sturdy tower which has—as in the case of so many other churches—come through the bombing essentially unharmed, dates from 1658-59, a time when very little church architecture was being built. One result of the bombing of the church was to bring to light a Saxon arch and the remains of a Saxon cross.

ST. OLAVE'S, HART STREET was a very small 15th century church containing many beautiful monuments, some of them brought from the demolished St. Benet Gracechurch. Samuel Pepys was a parishioner, and on the north wall of the chancel he erected a monument to his young wife, with a portrait bust whose gaze was directed to the Admiralty pew. The bust was removed before the bombing; its damaged setting is seen in the illustration.

ST. GILES'S, CRIPPLEGATE, is a Gothic church dating mostly from 1545. Most of the mediæval work survives, including the tower, with its picturesque upper stage of 1683, and the nave arcades seen in the picture. Many of the remarkably fine series of monuments are also intact. Restoration has robbed the exterior of all the interest it ever possessed, and the interior has, to a less extent, suffered in the same way. The church stands on the edge of the City's biggest " blitzed " area in a setting impressively recalling the London of October, 1666.

Churches

THE TEMPLE CHURCH, the older part of which—the " round "—was consecrated in 1185, is a building which has suffered such dreadful " restoration " that hardly one visible stone of it can be certified as older than the first half of the 19th century. The mischief was begun by Sir Robert Smirke in 1826, continued by James Savage, Sydney Smirke and Decimus Burton, and completed in 1842. Not only were the ancient carvings replaced with bad fakes, but numbers of memorials were destroyed and the remainder moved to the adjoining yard or fixed in the dark, almost inaccessible, triforium of the " round." Even the Templars' effigies were " restored " and moved to make a neat symmetrical group at the west end. The fire of 1941 over-reached the damage already effected by architects, annihilated the 17th and 18th century monuments, and reduced most of the Templars to dust. The print shows the church as the restorers left it : the photographs show the devastation of the " round."

ST. STEPHEN'S, COLEMAN STREET, by Sir Christopher Wren, dates from 1674-81. Its plan was a slightly irregular oblong, and the interior was not particularly memorable though it did contain a very grand communion table, supported on carved eagles, and one or two other good pieces of carving. The exterior was graceless and dull, suggesting that here was a case where Wren's instructions had been in more than usually general terms. At what point, one wonders, did his control end, and how much of the modelling of the church do we owe to one of his colleagues, or even to Marshall, the mason? The records leave us guessing.

ST. LAWRENCE, JEWRY, was the most costly of Wren's churches, and looked it. Its architectural arrangement was not particularly striking, but the organ-case and vestry were of fabulous magnificence. The organ, a Renatus Harris instrument, stood well out into the church, with dramatic effect. The vestry was panelled with some of the finest work associated with Grinling Gibbons's name, and had a rich plaster ceiling with a painted quatrefoil panel. Externally the church had only one "show" side, towards the east, where windows and niches were worked into a Corinthian colonnade, pedimented in the centre. The whole of this feature is intact.

Churches

ST. MARY-LE-BOW, by Christopher Wren, was built in 1670-80. The fame of St. Mary-le-Bow rests on three things: the crypt, the steeple and the bells. The crypt, Early Norman, has survived. So, fortunately, has the steeple. What the interior of St. Stephen's, Walbrook, is among Wren interiors, the tower of St. Mary-le-Bow is among his towers. No other is so varied, so animated, and so far from Classic poise and restraint. The bells are, of course, no longer the old Bow Bells of Whittington's time. They perished in the fire of 1666 and were replaced by new ones in 1758. After the splendour of the steeple the nave was disappointing —spacious but oddly incoherent and restless. It consisted of three bays with aisles, the centre arches being stretched uncomfortably into ellipses. East and west ends were identical—a contradiction, one cannot but feel, to what the function of a church prescribes. The same Corinthian half columns were found flanking the altar and the entrance, the same groups of three round-headed windows with three circular windows above were seen by those looking towards the altar and by those turning westward to leave the church. Below, the spire as seen from the nave and a view through the entrance hall towards Cheapside with some fragments of the bells visible in the foreground. In the picture on the facing page is seen the monument by Banks to Bishop Newton, 1782.

Churches

ST. ANDREW'S, HOLBORN, was built by Wren in 1686 and belonged to the St. James's, Piccadilly, class. The Gothic tower was recased nearly twenty years later, and is remarkable for the very carefully studied and rather effective Baroque treatment of the windows. The interior was heavily redecorated in Victorian times by S. S. Teulon, and rendered excessively dark, so that it is doubtful if this church will have left any clear impression of its beauties on contemporary minds.

ST. ANDREW-BY-THE-WARDROBE occupies a site which the formation of Queen Victoria Street rendered conspicuous and imposing. It was one of the later Wren churches, dating from 1692, and very straight-forward and economical in design. The monuments included some elaborate specimens of early nineteenth-century work. The interior has been burnt out, but the tower remains. The church was restored by Garner about 1875 and the railings and piers were added by Sir Banister Fletcher in 1902.

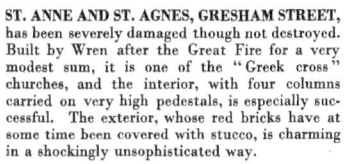

ST. ANNE AND ST. AGNES, GRESHAM STREET, has been severely damaged though not destroyed. Built by Wren after the Great Fire for a very modest sum, it is one of the "Greek cross" churches, and the interior, with four columns carried on very high pedestals, is especially successful. The exterior, whose red bricks have at some time been covered with stucco, is charming in a shockingly unsophisticated way.

ST. MARY'S, ALDERMANBURY, was built by Wren in 1667-70, probably on mediæval foundations. It is an unattractive building, not improved by the Italian tracery inserted, in spite of protests from the architectural press, by Victorian Mr. Woodthorpe. Wren was never at his best in small, aisled churches on the Gothic model, while the naïve detail of the east end suggests that his supervision here was not very close. The interior, with colonnades supporting a barrel vault, was more admirable than the exterior.

Churches

ST. BRIDE'S, FLEET STREET, one of the largest and costliest of Wren's churches, was built in 1670-84. It has been completely gutted, but the walls and arcades stand, in emaciated caricature. In this church Wren placed the galleries between columns rising from the floor, a more satisfactory device than the dwarf order rising from the gallery. This was a noble interior, a match for the famous spire, in which a difficult theme is so deftly handled as to seem without artifice or effort. Above, the still smouldering nave; on the facing page, the nave looking towards the east end.

Churches

ST. SWITHIN'S, CANNON STREET, with the alleged "milliarium" of the Roman roads recessed in its street front, was built by Wren between 1677 and 1687. The mason, as in the case of so many other City churches, was Joshua Marshall. Its interior was most interesting, with an eight-sided dome placed over a square and pierced by lunettes. The church had suffered from the absurd habit of Victorian restorers (in this case Woodthorpe) of inserting Florentine tracery in Wren's admittedly rather soulless round-headed windows. The tower had a very plain lead spire.

ST. MARY ABCHURCH, which has been slightly damaged, was finished by Wren in 1686. It is nearly square on plan and covered by an enormous dome, springing from arches which ride across the four corners in a daring and impressive fashion. The dome was painted by James Thornhill some considerable time after the completion of the church, and part of this painting is included in the damage. Another feature of this lovely building is the Baroque reredos, as successful in scale and placing as it is fresh and vigorous in design.

ST. STEPHEN'S, WALBROOK, the pride of English architecture and one of the few City churches in which the genius of Wren shines in full splendour, has been blasted but not destroyed. It can be, and must be, perfectly restored after the war. St. Stephen's, built in 1672-87, is, in a sense, Baroque; but the Baroque idea has floated into the still waters of Restoration England. There is as little emotion here as in Locke's *Essay* and no drama, except the fortuitous drama of changing light and shade. The columns, arches, spandrils and dome follow each other with the logic of a Euclidean theorem.

Churches

CHRISTCHURCH, NEWGATE STREET, had a most interesting and successful gallery arrangement, the gallery fronts being supported on the pedestals of the columns. The galleries were thus set rather low in the church, giving the interior a noble air, somewhat spoiled by the building up of the rear part of the galleries against the windows. The vault and clerestory were also treated in an unusual way. The finest part of the church, however, was the tower, which still stands, though badly burnt. It was the most mature and irreproachable of all Wren's steeples and anticipated in a curious way the purified picturesqueness which the Greek revivalists sought, but which few of them were clever enough to achieve. The church was built between 1677 and 1691, but the spire was not finished till 1704.

ST. VEDAST'S, FOSTER LANE, was built by Wren in 1670-73 and the brilliant and magnificent Baroque tower added many years later. The interior consisted of a spacious nave with a flat plaster ceiling, and a single aisle, divided from the nave by the elegant arcade seen in the left-hand picture. This was one of the most successful, though one of the cheapest, of Wren's " hall " churches. It has been badly scorched, most of the woodwork having gone, but the structure as a whole is not beyond reconstruction, and having regard to the special merits of the tower it may be hoped that this will be done. The right-hand picture shows the wrecked interior looking towards the west end.

ST. NICHOLAS COLE ABBEY, designed by Wren and built in 1671-81, was internally a simple Corinthian-pilastered hall, with three arches at the west end filled with a gallery. Much alteration had been done in 1873 but the church was full of good fittings, with enormous eighteenth century monuments on the walls. It has been completely burnt out, and the tower has lost its quaint octagon spire with the tiny balcony slipped on its apex like a ring.

Churches

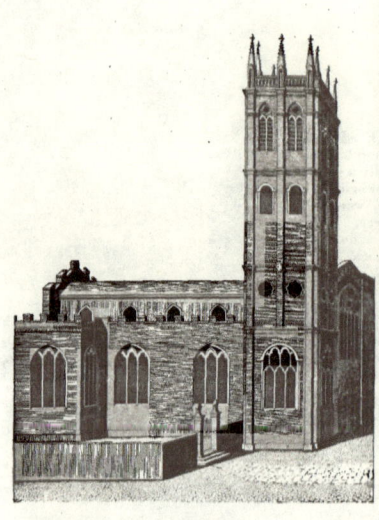

ST. ALBAN'S, WOOD STREET, which has been gutted, although the tower stands, is a remarkable example of the persistence of Gothic at the end of the seventeenth century. It was rebuilt in Gothic in 1634 and again in the same style after it had been burnt down in 1666. The tower is certainly from drawings by Wren, and may be described alternatively as late Gothic survival or early Gothic revival. The church was built in 1682-7; the apse and other details were added by Sir Gilbert Scott in 1856. The contents were of little importance.

ST. JAMES'S, PICCADILLY, built by Wren in 1682-4, and therefore somewhat later than most of the City churches, was considered by its architect to present the most satisfactory solution for a galleried church to hold a large congregation, all of whom could see and hear the preacher. Though badly damaged, it still stands and is capable of reconstruction. Most of the fittings have been saved, including the altar-piece, one of the few authenticated works of Grinling Gibbons. The remarkably ugly spire which has disappeared from the tower was designed, not by Wren, but by a Mr. Wilcox, a carpenter, who undercut Wren's estimate. Wren at one time intended a domical termination, covered with a scaly design, probably in copper plates. The early Victorian rectory and the gateway designed recently by Sir Reginald Blomfield have both gone, and provide an opportunity not to be missed when the time for reconstruction comes. The outside pulpit, conspicuous on the Piccadilly façade, was by Temple Moore, 1902.

Churches

ST. AUGUSTINE'S, WATLING STREET, has lost its fantastic lead-covered spire, and the church itself has been partly destroyed. The photograph, taken across the ruins of surrounding buildings, shows its situation in the shadow of St. Paul's. Finished by Sir Christopher Wren in 1683, the interior suffered from modernization by Sir Arthur Blomfield in 1878 when the panels of the barrel vault were ornamentally glazed. Arcades and vaults sprang from Ionic columns, pitched very high, in order to preserve their proportion, on thin panelled pedestals. When the high old pews were removed the effect of this became grotesquely exaggerated. The tower, whose masonry details are particularly sensitive and beautiful, was finished some years after the church itself.

ST. MILDRED'S, BREAD STREET, was a simple domed church of great beauty, finished by Wren in 1683. Deep arches with well-designed plaster-work supported the dome at east and west, and there was a particularly gorgeous pulpit and sounding-board. This was one of the least altered and one of the most eloquent and charming of all the City churches. Its destruction has been almost complete.

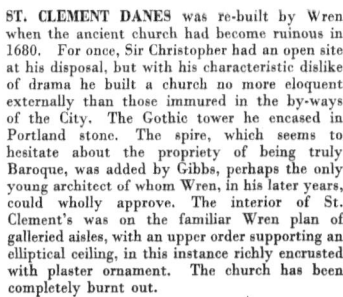

ST. CLEMENT DANES was re-built by Wren when the ancient church had become ruinous in 1680. For once, Sir Christopher had an open site at his disposal, but with his characteristic dislike of drama he built a church no more eloquent externally than those immured in the by-ways of the City. The Gothic tower he encased in Portland stone. The spire, which seems to hesitate about the propriety of being truly Baroque, was added by Gibbs, perhaps the only young architect of whom Wren, in his later years, could wholly approve. The interior of St. Clement's was on the familiar Wren plan of galleried aisles, with an upper order supporting an elliptical ceiling, in this instance richly encrusted with plaster ornament. The church has been completely burnt out.

Churches

ST. ANNE'S, SOHO, was consecrated in 1686. Its designer's name is unrecorded but it is pretty safe to guess that he was one of the many masons whom Wren had been employing in the City. It was a plain, galleried church with an apse flanked by square chambers—a strange revival of one of the earliest forms of Christian church architecture. The tower, which survives, is a curious affair built in 1802-6 and designed, apparently, by S. P. Cockerell. It has much of the spirit of Soane's modernism, and in any city but London might even be supposed to be a rash adventure in *art nouveau* or *jugendstil*.

ST. NICHOLAS, DEPTFORD, is a little-known church, interesting because of its monuments to sailors, adventurers, ship-builders and members of the Evelyn family. There is a mouldering Gothic tower at the west end, but the church itself was rebuilt in 1697. Within twenty years it showed signs of failure and the wrought-iron braces seen in the illustration were inserted to hold the structure together. The rich "overthrows," intended to disguise these necessary makeshifts, give a curious effect. All the monuments have escaped serious damage and many have now been removed, together with the pulpit, reredos and organ case, all of which are good seventeenth-century work.

ST. ALPHEGE'S, GREENWICH, was a particularly splendid church, by Nicholas Hawksmore, built in 1714 under Queen Anne's Act for Building Fifty New Churches. Happily, the immense solidity of the construction has resulted in the shell remaining intact, but much fine woodwork was lost in the fire. The arrangement of the galleries in this church was particularly successful. The only disappointing part of the building is the western spire, which was not part of the original design but a poor imitation of Gibbs by John James.

Churches

ST. GEORGE'S-IN-THE-EAST was, and partly remains, one of the grandest works of Nicholas Hawksmore. It survives as a majestic shell, the interior scorched and vacant, the exterior hardly even scarred. It was built in 1715-25 and was one of the three churches built by Queen Anne's Commission in the old, vast parish of Stepney. Both in plan and general handling St. George's was profoundly original. It was a Greek cross church with an apse at one end and a tower, containing a vestibule, at the other. Four smaller staircase towers soared up into octagon turrets crowned by lead cupolas. There was something almost Elizabethan about these strange, romantic steeples, giving a skyline which would have been more like a castle than a church but for the dominating mass of the western tower—a "Boston stump" as a Roman might have conceived it. The interior was less emotional and, indeed, rather cold. But the gallery fronts and doorcases were finely carved and the pulpit was a lovely piece of joinery, with marquetry panels. Few alterations had been made, but delightful coloured glass had been placed in the apse windows in the early nineteenth century.

34

Churches

ST. JOHN THE EVANGELIST, SMITH SQUARE, WESTMINSTER, is a building of the highest importance in the history of our architecture. Hardly anywhere else in England has the Baroque manner been so dramatically and forcefully handled. It was one of the churches built under Queen Anne's Commission, was designed by Thomas Archer and built in 1721-8. In two respects Archer's church was unlucky. First, because lead cupolas were placed on the towers, instead of the exciting Baroque pinnacles he designed; and second, because the church was gutted by fire during the eighteenth century and the interior tamely reconstructed in 1758. Its most recent misfortune has, in fact, involved none of Archer's work and the shell must be regarded as a very precious survival.

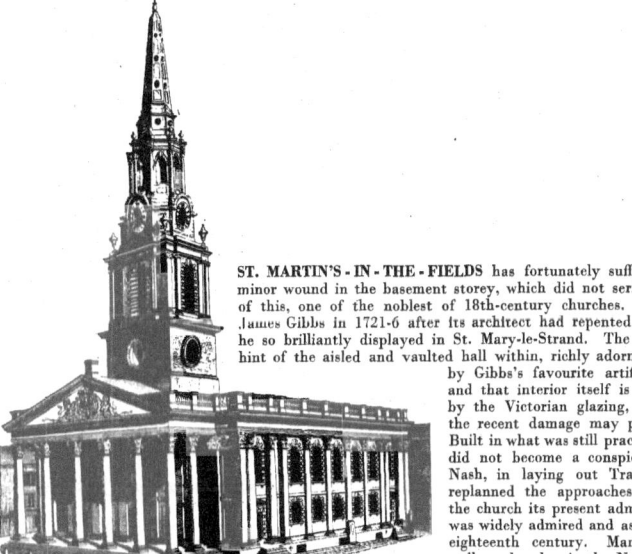

ST. MARTIN'S - IN - THE - FIELDS has fortunately suffered nothing worse than a minor wound in the basement storey, which did not seriously disturb the structure of this, one of the noblest of 18th-century churches. St. Martin's was built by James Gibbs in 1721-6 after its architect had repented of the Baroque tendencies he so brilliantly displayed in St. Mary-le-Strand. The serene exterior gives little hint of the aisled and vaulted hall within, richly adorned with plaster arabesques by Gibbs's favourite artificers, Artari and Bagutti; and that interior itself is robbed of half its beauty by the Victorian glazing, for the removal of which the recent damage may provide a plausible excuse. Built in what was still practically a village, St. Martin's did not become a conspicuous public building until Nash, in laying out Trafalgar Square, ingeniously replanned the approaches to the Strand and gave the church its present admirable setting. The church was widely admired and as widely imitated during the eighteenth century. Many provincial churches, as well as churches in the New World, echo its plan and silhouette (see, for example, the Liverpool church on page 113).

Churches

ST. JOHN'S, HORSELYDOWN, is in Bermondsey. The name "Horselydown" is a designation surviving from the time when the church stood among fields. The most conspicuous thing about it, in every sense, is the steeple. There is an excellent view of it from the train, a few hundred yards out of London Bridge Station, and probably every architect has, at some time or another, found his gaze

riveted to the broad-bottomed Ionic column which serves as a spire. This wild vagary, attached to a church of supreme dullness, is the most improper thing in the whole of English architecture. Very little is known about the church, except that it was built in 1727. The architect is said to have been John James of Greenwich, but this may be a libel on that reasonably competent official architect. Quite apart from the " spire," nothing could be clumsier than the way its pedestal juts up between the two halves of a broken pediment. The rest of the church suggests a passing acquaintance with the work of Hawksmore, but the relationship of the several types of window is maladroit, and altogether the building is a pretty exemplar of what happens to Palladian architects who do not go by the rules.

Churches

ST. BARNABAS, KING SQUARE, FINSBURY, was designed by Thomas Hardwick, the surveyor to the St. Bartholomew's Hospital estate on which King Square stands. It is one of the churches built under the Church-building Act of 1818 and was begun in 1822. This was the same year as All Souls, Langham Place (page 43), so it is impossible to guess whether Hardwick or Nash was first with the idea of a short, needle-pointed spire raised over an Ionic portico. Hardwick had already built several London churches, including St. John's Wood and St. Marylebone. He was an old man in 1822 and, as a pupil of Chambers, had always held conservative views; so the character of the King Square church may well be due to his son Philip, of Euston Arch fame. The badly wrecked interior of the church was of the plainest possible type.

MARY'S, ISLINGTON, was built the site of an earlier church in the dle of the eighteenth century, when gton was one of the main centres uburban development and the place residence of prosperous city men. architect was Lancelot Dowbiggin, rpenter, who built a plain church a steeple of considerable richness, iously inspired by the work of the r Dance. The roof of the nave was e piece of carpentry, and an engrav- of it appears in one of the con- porary carpenters' text-books. Al- gh the church received a direct hit, tower and spire still stand, rently undamaged, and, it may be d, will remain.

HOLY TRINITY, MINORIES, was a modest squarish church, built in 1707 and later stuccoed. It had a Gothic predecessor which incorporated some remains of the monastic church of the Minorites, founded in the twelfth century. Some of the medieval masonry is still there and was revealed when the church was burnt. It is shown in the photograph alongside.

Churches

ST. JOHN'S, WATERLOO ROAD, one of the State-built "Waterloo" churches, was designed by Francis Octavius Bedford and erected in 1823-4. Its Greek Doric portico and the timid little steeple behind it still stand. The interior of the church, a very plain hall with some notable modern fittings by Mr. J. N. Comper, has been practically destroyed. The illustration shows some characteristic details of this very bare product of Greek revivalism, whose portico, nevertheless, will be missed from the conspicuous place it occupies near a railway terminus and a river crossing.

ST. MARY'S, HAGGERSTON, built in 1826-7, was one of the large number of churches erected in the first half of the nineteenth century by the Commissioners for Building New Churches. Nash, by virtue of his position on the Board of Works, was one of the Commissioners' architects and condescended to supply a few designs himself. St. Mary's was a plain brick box with a rather elaborate Tudor front dominated by an absurdly thin tower with an even thinner lantern on top of it. The erection of the church was supervised by John Walters. St. Mary's was remodelled in Victorian times by James Brooks, who somewhat modified its box-like austerity. The church has been reduced to the great heap of rubbish seen in the picture, even the tower being completely destroyed.

ALL SOULS, LANGHAM PLACE, is a Commissioners' Church and was built in 1822-4. Its plan, contrived in such a way as to provide a terminal feature at the end of Regent Street although the church itself lies at a different angle, is one of John Nash's most brilliant conjuring tricks. The two peristyles and the sharp spire are combined with astonishing success, the one questionable feature being the continuation of the rake of the spire to the base of the surrounding columns. The church was much ridiculed when it was new, a Member of Parliament stating in the House that " he would give a trifle " to have it pulled down. This was the occasion for the famous caricature of Nash spitted on his own sharp steeple, with the caption " *Nash*ional Taste ! ! ! " As the photograph shows, the only considerable exterior damage is the removal of the tip of the spire.

Churches

ST. MARK'S, KENNINGTON, is, like St. John's, Waterloo Road (page 42), one of the four churches in South London dedicated to the four Evangelists and intended to serve the needs of the enormously increasing population of the parish of Lambeth. These churches were paid for by the Church Commissioners and were among their most ambitious architectural efforts. For some reason they became known as the "Waterloo Churches," though this title was never applied to them when they were new and was possibly coined when the "Strand Bridge" became Waterloo Bridge. The Kennington church was built in 1822-4, the ostensible architect being D. R. Roper, a successful surveyor. Roper, however, employed A. B. Clayton, a "ghost" whose clever work is represented elsewhere under other men's names.

ST. JAMES'S, WEST HACKNEY, stands in the Stoke Newington Road and was built in 1823-5 by Robert Smirke, an architect who was never afraid of repeating himself and always did so with accuracy and skill. Many things could be less elegant but few duller than this typical Smirke church, with its efficiently executed details and marvellous want of imagination. Not even a direct hit could ruffle its utter complacency.

ST. GEORGE'S CATHEDRAL, SOUTHWARK, was built in 1840-48 by A. W. Pugin, and though it had not the delicate charm of his churches at Birmingham and Derby it was, and in part remains, a beautifully conceived piece of Gothic. Pugin's architecture is always rather thin, partly, no doubt, because he " starved his roof-tree " for lack of funds, but also because, as all his sketches show, he liked thin architecture. St. George's had its faults: the overcrowded tracery and the overcrocketted pinnacles. But it had, too, sensitively moulded arcades, lovely glass, and altar-pieces which only a decorative genius of a high order could have produced. Pugin designed a grand tower and spire for this church, which, had they been built, would have redeemed the rather sad inconsequence of the exterior. The church has been entirely gutted, but the arcades, exposed to full daylight, are seen to great advantage, and the specially beautiful altar-piece in the north chapel is unharmed.

Churches

ST. ALBAN'S, HOLBORN, was built in 1859-63 and was one of the principal works of William Butterfield. It has been completely burnt out and the stonework much damaged, but in the ruin there is still everything of the forceful, eccentric spirit of its architect. The high western arch opening into the saddle-back tower, with two great windows beyond, was perhaps the grandest part of the church, and those who knew it before the fire will remember Mr. Comper's spired font-cover, a fountain of golden pinnacles soaring up towards the arch. By Mr. Comper, too, was the Stanton chantry in the south aisle. The east wall is covered with water-glass paintings by Le Strange. The original reredos, modest and angular, was covered by a great new reredos, of 15th century type, by Bodley and Garner, which has been quite destroyed.

ST. CLEMENT'S, CITY ROAD, was one of William Butterfield's most memorable churches. W. R. Lethaby used to draw a line between the " hard " and the " soft " Gothic Revivalists. Butterfield was as hard as nails. With merciless consistency he used the vernacular materials of his time—pit-sawn timbers, harsh red and brown bricks, shiny tiles—and out of these he made a rugged poetry, more energetic and moving than any that the Gothic Revival produced. The photograph shows the wreck of St. Clement's before the " west " wall (it stood, actually, north and south) was demolished. Built in 1874, some years after St. Alban's, Holborn (page 46), and nearly thirty years after All Saints, Margaret Street, St. Clement's was lofty and severe, with a single aisle and a high lancet-pierced clerestory. The bell-turret, with its original and exciting pattern of brick and stone, the high-pitched east window and the angular waggon roof were all typically Butterfieldian. The tracery and mouldings were, as usual, strictly according to medieval precedent; it was the disposition and proportioning of the church which gave it its strong appeal. Its loss, together with the loss of St. Alban's, Holborn, is all the more unfortunate in that the work of Butterfield has still not had the study and recognition it deserves.

Churches

ST. JOHN THE DIVINE, KENNINGTON, was one of the last and principal churches of George Edmund Street and was built in 1870-4. In this, as in other churches of his, Street was concerned with providing a very broad nave, from which all the congregation could see the altar. The aisles he used merely as passages, canting the easternmost bays of his arcades inwards to join the chancel arch. The nave he covered with an immense pointed barrel vault in timber. It is worth comparing St. John's with another broadnaved church by Street which has been burnt, All Saints, Clifton (page 89). Externally, St. John's was remarkable for its tower and broach spire, designed by the architect's son on the basis of Street's own intentions. The interior was richly furnished and included a reredos by Bodley and Garner.

ST. JOHN'S, RED LION SQUARE, has always been recognised as J. L. Pearson's masterpiece; and a masterpiece it certainly was. Built in 1874-8 on a cramped and irregular corner site, it displayed a mastery of the art of stone vaulting never seen since the Middle Ages. There was a short, broad nave whose arcades returned on north and south to meet the chancel arch. The aisles were of unequal length and width; but everywhere the vault system ordained graceful unity, most subtly of all, perhaps, where the narrow south aisle opened into a chapel, filling, with the tower, an awkward elbow of the site so that no awkwardness was apparent. The detail, throughout, was a gracious, sober Early English, rather too dainty for some tastes, but technically perfect. On the north of the church, the parsonage survives, a pleasant building which helped to bring the soaring lines of the church into harmony with the Square. On the south is the stump of the intended tower which was to have risen into a three-stage belfry and a handsome spire of the Normandy kind. St. John's has been blown to pieces, and since that event ruthless demolition has taken place. But Pearson's drawings exist, and it would hardly be mistaken piety to bring this church back to its former perfection.

Churches

THE CHAPEL OF THE ASCENSION, BAYSWATER ROAD, belongs to one of the art-movements of the 'nineties. Its architect, Herbert P. Horne, was a poet and illustrator of talent and a frequent contributor to the "Hobby-horse," along with Selwyn Image and others. The Chapel was built in 1890-94 after Horne had spent a holiday in Verona and Assisi with his partner A. H. Mackmurdo. The walls of the interior were adorned with paintings by Frederick Shields. It has been completely gutted.

The Church of OUR LADY OF VICTORIES, KENSINGTON, which became the "pro-Cathedral" of London, was built in 1867 by George Goldie, the Roman Catholic architect, whose chief work is St. James's, Spanish Place. A very high building, richly carved in the "early French" taste, it represented, to some extent, the influence of Cardinal Manning in overcoming Catholic prejudice against the Gothic Revival, with its decidedly Anglican associations.

THE METROPOLITAN TABERNACLE (colloquially, Spurgeon's Tabernacle), in Newington Butts, was an iron structure within a masonry shell to which a clumsy Corinthian portico was attached. It was built in 1898-99, the architects being Searle and Hayes. Its predecessor, destroyed by fire, was an even more capacious building erected in 1861 for the popular preacher, Charles Haddon Spurgeon.

50

THE CITY TEMPLE, Holborn Viaduct, was London's largest and best known Congregational Church. Consecrated in 1874, it was basilican in form and Italian in character, evincing that lack of sensitiveness which is so invariable in Victorian non-conformist architecture as to seem deliberate. The piers of nave and galleries, as can be seen from the photograph above, are of cast iron. The church, which held a congregation of 3,000, was designed by Lockwood and Mawson, north-country architects whose work, in its natural habitat, has a certain grim grandeur. An early work by Lockwood is the burnt-out Congregational Chapel at Hull, illustrated on page 124. In the photograph on the facing page the tower is seen from within the burnt-out church.

Westminster

THE PALACE OF WESTMINSTER, designed by Charles Barry with the copious assistance of A. W. Pugin, is the successor of a conglomerate of buildings, partly medieval, with additions by Kent, Wyatt, Soane and others, which was burnt to the ground in 1834, with the exception of Westminster Hall and the crypt of St. Stephen's Chapel. The Hall, which still survives, was built by Richard II, one of the few English medieval monarchs to be greatly interested in the arts. Its roof is a miracle of Gothic carpentry, easily the most remarkable timber structure of its age in any country. Fortunately, the damage it has suffered recently is slight. The Hall was brilliantly incorporated by Charles Barry in his great axial plan for the new Palace, begun in 1837. Most of the working drawings passed through Pugin's hands; and the combination of Barry's feeling for broad masses and Pugin's decorative genius resulted in a fine building whose reputation has suffered more from Victorian prejudice against " late " Gothic than from any shortcomings of its own. The Commons chamber (large picture on the facing page) which has been burnt out, was much spoiled by the lowering of the ceiling to the level of the oblong openings seen in the photographs. The two small pictures alongside show one of the court-yards and the " aye " lobby.

53

Westminster

THE HOUSE OF COMMONS was originally lit by ranges of tall windows like the one seen in this picture. But when the ceiling was lowered to improve the acoustics the upper lights were cut off and the architectural effect of the chamber spoiled. Here the windows are seen once more in their original proportion, calcined into a rather beautiful semblance of hoary antiquity. They contained stained glass from Pugin's designs.

WESTMINSTER ABBEY, though seriously damaged at the crossing, has suffered less than the complex group of buildings which surrounds it. These include the burnt-out Deanery, lying alongside and over the west walk of the cloister; also the little cloister with its charming residences, originally almshouses but rebuilt in the seventeenth and eighteenth centuries; also the buildings of Westminster School. The School Hall, with its hammerbeam roof, dating from the seventeenth century, has been gutted, and so, unfortunately, has Dr. Busby's library. Left, debris on the floor of the Abbey; on the facing page, Dr. Busby's library and adjoining buildings, and the burnt-out Deanery from the cloisters. Above, the dormitory of **WESTMINSTER SCHOOL.** Though burnt out from end to end it is externally intact and the long façade to the Dean's garden looks as noble as ever. Wren's name has often been mentioned in connection with the building, but in fact it was designed by the Earl of Burlington and begun in 1722. It was the first structure in London to be faced with Bath stone, preceding the usual claimant to this distinction, Gibbs's gateway at St. Bartholomew's, by eight years. Little of architectural value was lost in the burning of the interior, and it is hoped that the shell may be found capable of re-use.

Charterhouse

THE CHARTERHOUSE has lost many of its buildings, though the walls stand for the most part and the chapel, with its very precious screen and monuments, is quite unharmed. The Hall has suffered badly. It was built as part of the great mansion of the Norths, after the dissolution, and subsequently incorporated in the hospital endowed by Sir Thomas Sutton in 1611. The great chimney-piece of that date and some of the woodwork have survived. The Master's House and adjacent buildings have been burnt out, and among the ruins lie carved stones which must have belonged to the Carthusian priory and were used up in the walls of the Norths' mansion. Among other buildings gutted by fire is the Hall of Merchant Taylors School, built in 1872, when Charterhouse School moved to Godalming and the Merchant Taylors migrated from the City to Charterhouse, which they left a few years ago.

The Tower

THE TOWER OF LONDON suffered a direct hit on the North Bastion. This massive and featureless structure had been built out from an angle of Edward I's outer curtain wall some time after 1842, when Anthony Salvin began his re-mediævalisation of the Tower. It was of no architectural consequence and should, in fact, never have been built. The photograph shows the extent of the damage, with the towers of the inner curtain wall, the Victorian barracks and one turret of the White Tower. Nothing of historic value in the Tower has yet been damaged.

Inns of Court

SERJEANT'S INN, off Fleet Street, was for centuries the home of the Honourable Society of Judges and Serjeants-at-law. Towards the end of the eighteenth century, however, most of the property was rebuilt as private houses, among which No. 13, shown in the engraving and now demolished to first floor level, was conspicuous for its fine Ionic façade which dominated this pleasant backwater.

STONE BUILDINGS, LINCOLN'S INN, built by Sir Robert Taylor in 1774-8 and comprising sets of chambers as well as the principal offices of the Inn, represent part of an ambitious scheme of rebuilding, only half realized. Taylor was chosen for this work after Adam, Paine and Brettingham had submitted drawings, his drawings and theirs being still in existence. Like all Taylor's work, Stone Buildings are proportioned and detailed in such a way as to give a weighty, imposing effect. Carefully considered profiles and a liberal introduction of string-courses are responsible for this, and it is interesting to compare this building with Taylor's house in Lincoln's Inn Fields, page 72. Stone Buildings still bear the marks of a bomb which fell during the last war.

Inns of Court

AY'S INN dates chiefly from the last
rter of the seventeenth century, when
chambers were rebuilt after successive
s. Both the Hall and the Chapel,
vever, incorporated earlier work.
e Hall, left, was reconstructed about
6-59, with a very fine hammer-beam
f curiously enriched with classical
ail. In Elizabethan times this was
ene of festive life often visited by the
een herself, and the fantastic oak
en was said to be her gift to the Inn.
re the *Comedy of Errors* was produced
1590. The exterior of the Hall was
newhat altered at the same time that
 adjoining library was built, in
9-41, the Inn's surveyor at the time
ng Francis Wigg. Later alterations
e made by Isaacs and Florence. Both
l and Library have been gutted; and
whole of the fine woodwork is lost.
 Chapel (small picture), a small build-
 of very minor interest, had been
ovated some years before the war by
 Edwin Cooper. The chambers,
ugh not among the most scholarly
ductions of their time, were fine
cious apartments where centuries of
essional men had found it a pleasure
 live and work. Among architects,
*. Bodley and Sir Giles Scott have had
es in the Inn at one time or another.
 lower photograph on the facing page
vs a corner of Gray's Inn Square.

THE TEMPLE as a whole has been extensively
damaged in several raids. The devastation at the
Temple Church, which occurred in a subsequent raid
to the one which caused the damage shown, is
illustrated on page 16, and the damage to Middle
Temple Hall on pages 60 and 61. The photographs
on the right show (top) Crown Office Row, (centre)
Middle Temple Lane, looking south, and (bottom)
Pump Court. Of these, Pump Court was the finest,
consisting of excellent red brick chambers, erected
after a fire in 1678. Crown Office Row comprised a
brown brick building by Timbrell, a carpenter, 1737,
and a florid block by Sydney Smirke, 1853-4, both
quite destroyed. Middle Temple Lane has lost some
seventeenth-century buildings by Barbon, a large
block by Sir Robert Smirke and Victorian chambers
by St. Aubyn.

Middle Temple Hall

MIDDLE TEMPLE HALL has sustained severe damage to its eastern end, where part of the wall and roof have been demolished and the screen wrecked, though not destroyed. Built in 1562-72, the Hall has a massive hammer-beam roof, whose heavily moulded pendents can be seen in the view on the opposite page. Shakespeare is said to have taken part in a performance of " Twelfth Night " in this Hall. Externally, the Hall was considerably altered by Henry Hakewill and, after him, James Savage, round about 1830.

Trinity House

TRINITY HOUSE, which faces the north side of the Tower of London across Tower Hill, was built in 1793-95 under the direction of Samuel Wyatt. The character of the design, however, has sometimes given rise to the suggestion that Samuel's more famous brother, James Wyatt, was concerned in the work. The stressing of horizontal lines and the Græco-Roman details are strongly reminiscent of James's style. The interior, which has been entirely destroyed, along with its rich contents and ships' models, had a delightful entrance and staircase hall and some well-decorated rooms. This building was to occupy an important place in the Tower Hill improvement scheme, and it is to be hoped that it will be regarded as one of those buildings where reconstruction of the original design may be properly undertaken.

Somerset House

SOMERSET HOUSE was England's greatest public work of the eighteenth century. It was consciously built as such by an architect whose aim was to set English architecture on the level of aristocratic repute which it enjoyed in France and Italy. Sir William Chambers, the one great eclectic master of the English school, began Somerset House in 1776. It was built to accommodate several public offices and learned societies, the block towards the river containing the Navy Office, or that part of it which was not housed in Whitehall. It is this portion of the building—England's nearest parallel to the vast Admiralty façade on the Neva at Leningrad—which has received the most serious damage. The engraving shows the building before the formation of the Embankment, with the water-gates opening on the river.

Buckingham Palace

BUCKINGHAM PALACE has sustained damage in several parts, including the Chapel, and the Garden Entry seen in the accompanying illustration. This is part of Nash's design, but was completed under Edward Blore after George IV's death and Nash's deposition. It forms part of a wing stretching north from the main block. The materials include Bath stone, cast-iron for the Doric columns and Coade's patent stone for the Royal arms. This is one of the few parts of Nash's original building visible to the public.

Holland House

HOLLAND HOUSE, a Jacobean mansion standing within its own park, in the heart of London, is one of the capital's most curious and beautiful survivals. Part of it is the work of John Thorpe, authenticated by a drawing in the Soane Museum. It was built for Sir Walter Cope as "Cope Castle" and altered at various periods after his time, first by Lord Holland in 1638-40 and again in 1704, 1748, 1796 and during the nineteenth century, when the revived taste for Jacobean art led to some unfortunate faking. The library, which has been wrecked, is in one of Thorpe's wings and contains two original fireplaces of good design; the destroyed upper part was of little importance. The principal staircase belongs to the earliest part of the house and is an excellent example of its kind. During the eighteenth and early nineteenth centuries Holland House became the great centre of Whig society, and its history during those times has been well written by the present owner, Lord Ilchester. The large picture on the facing page shows the library and the small ones (below) show the door of the burnt-out breakfast room, the grand staircase and wrecked furniture.

Kensington Palace

KENSINGTON PALACE consists of a main block and other buildings by Wren, and an east wing, containing some of the state rooms, by William Kent. The damage has been confined to the upper storeys, as shown in the photograph, in the background of which is seen the cupola of the stable block, dating from 1691 and designed by Wren.

Chelsea Hospital

CHELSEA HOSPITAL has suffered considerable damage in part. Built in 1682-92, by Wren, it is one of the first of Sir Christopher's buildings to introduce the broad simplicity and big scale of his later manner. The stone cupola over the Doric portico is particularly admirable. The photograph shows where a bomb has destroyed the organist's rooms, once occupied by the famous Dr. Burney. Damage was also inflicted on Soane's infirmary buildings.

Clubs

THE NAVAL AND MILITARY CLUB (colloquially the "In-and-Out"), is a mid-eighteenth century mansion known successively as Egremont House, Cholmondeley House and Cambridge House. Palmerston lived here for a time, and the club was inaugurated here in 1862. With the St. James's Club and Burlington House it is among the last of the great Piccadilly mansions of the eighteenth century. The architect is unknown, but the name of Vardy, Kent's associate, may be suggested.

THE ARTS CLUB, Dover Street, Piccadilly, an eighteenth-century building, later altered inside, was taken over by the Club from Lord Stanley of Alderley in 1896. Its plan proved remarkably well suited to a club of the intimate kind (as distinct from the monumental kind represented by the great Pall Mall clubs). The suite of rooms comprising the first floor was a very good example of the London town-mansion interior equally suitable for formal or domestic occasions. The exterior was similarly characteristic of the London street architecture of its time.

Domestic

HARPUR STREET (top, and lower left) was formed in the seventeenth century and rebuilt about 1760. It was perfectly symmetrical and uniform in detail. The east side has been completely destroyed; of the west side a little more than half remains, including the pedimented centre house (No. 16) which has an interesting interior. From 1767, this was the home of Dr. John Fothergill, political associate of Benjamin Franklin, and the first physician to diagnose diphtheria. The street runs north from Theobalds Road on the estate of Bedford College, from whose founder, Sir William Harpur, it takes its name. The neat joinery of the Doric doors and the precise spacing of the windows made this a distinguished street. The false pediments on the centre houses were typical of the period and its tendency to artificial over-emphasis. **FEATHERSTONE BUILDINGS** (lower right) was a narrow street north of Holborn, built about 1720 and, until the bombing, almost intact. A few houses on the west side precariously survive, with their carved hoods and panelled rooms.

Domestic

The west side of **FINSBURY SQUARE**, now almost entirely demolished after being burnt out, was a symmetrical block designed by James Peacock, an assistant of the younger Dance and an able architect. He designed the first Stock Exchange and published a quaint book of domestic plans under the anagrammatic pseudonym of " Jose MacPacke " (described on the title-page as a bricklayer's labourer). Built about 1777, the originality of composition and skilful interpretation of the Adam manner in these houses, and their frank, logical, roof treatment, have never been fully recognized.

MECKLENBURGH SQUARE. The east side of this square was the only part of the Foundling Hospital estate in which the Governors' original architectural ambitions were realized. It was built around the year 1812, the design being exhibited in the Royal Academy of that year by a little known architect, Joseph Kay, whose only other works of consequence are the street improvements at Greenwich and the colonnades on either side of the Queen's House. Kay was a pupil of S. P. Cockerell, succeeding him as Surveyor to the Foundling estate in 1807. He laid out the gardens in the middle of the square and gave great satisfaction to his employers. He lived in Gower Street, was actively interested in the formation of the R.I.B.A., and was buried, at the Governors' request, in the Chapel of the Foundling Hospital. The other façades in the Square are not from Kay's design.

PORTMAN HOUSE, otherwise known as Montagu House or simply as No. 22 Portman Square, possessed some of the finest domestic interiors in London. The house lies across the north-west corner of the Square, in its own grounds. It has been entirely burnt out, and only the shell, which had undergone considerable modernization, survives. It was begun some time after 1760 for Mrs. Elizabeth Montagu, the queen of intellectual lion-tamers, for the express purpose of providing a centre for high-brow parties. Her architect was James Stuart, the pioneer of the Greek Revival, whose exquisite, sparkling decorations, in which he received the help of Bonomi, formed the chief beauty of the house. The work proceeded spasmodically and the building was still unfinished in 1781. It was here that Mrs. Montagu gave her annual May-day parties for chimney-sweeps. The house subsequently reverted to the Portman family, whose London residence it was until its destruction. The illustration shows the front drawing-room on the first floor, with the surviving indications of its segmental ceiling. The Athenian capital, seen through the gap in the floor, reminds us of the novel architectural scholarship which made Stuart famous.

Domestic

SUFFOLK STREET, near the Haymarket, was laid out by Nash about 1820, nearly on the lines of an older street. The buildings, by several architects (Nash included), were completed within the next few years. The illustration shows the north end with the remains of Garland's Hotel, whose quiet, comfortable, old-fashioned interior was "discovered" by many an American in search of traditional England. The building on the left, with the fan-headed window (a favourite Nash motif) is part of the back elevation of the Haymarket Theatre.

PARK CRESCENT was begun in 1812 as part of Nash's first far-reaching plan for the Regent's Park. It was meant to be the south half of a vast circus. Six houses were ready in 1819, the rest in 1822. Foundations were laid for some of the northern houses, but the scheme was altered in favour of an open vista through Park Square. The Crescent is nicely proportioned and detailed, in contrast to the slap-dash scenic designing of the later terraces.

70

CARLTON HOUSE TERRACE has been battered throughout the whole length of both blocks, exposing the "shoddy" construction of brick and stucco in a way that would have delighted Victorian critics of John Nash. The terrace dates from 1827-9, when it was built on the site of Carlton House and its gardens. Various architects designed the individual houses; Nash leased and designed a few. The ornament in the pediments is by Bernasconi. Controversy raged round the question of the Terrace's preservation in 1933, when rebuilding was begun to the design of Sir Reginald Blomfield. Both the merits and demerits of the terrace were grossly exaggerated. It possesses both, but the grandiloquence of the general conception is esteemed by many to compensate for the impudently slap-dash technique.

Domestic

THE PARK VILLAGES, EAST AND WEST, were the last parts of the Regent's Park layout to be executed. Nash was always fond of the idea of a romantic suburban village but was prevented from following out all his intentions and was, in any case, too old to design more than a few of the houses. Most of the work was done by James Pennethorne. The illustrations show Park Village East, which contains one or two Gothic and Swiss curiosities. One of a pair of Gothic houses (No. 18) has been destroyed, but most of the others are intact. It would be difficult to exaggerate the beauty which this water-side village once possessed, but is unlikely to possess again.

NO. 35 LINCOLN'S INN FIELDS was one of the most impressive late Palladian houses in London. With No. 36 (long since rebuilt) it was designed by Sir Robert Taylor in 1754. The façade is unique among London houses in its suggestion of a super-imposed order by means of a double tier of string courses. The interior contained a splendid staircase and some interesting rooms, while the windows were glazed in a curious and rather effective pattern of hexagons. The first occupant of the house was a rich Georgian lawyer; the last was the College of Estate Management.

GUILFORD STREET and **GUILFORD PLACE** once formed a fine town-planning unit opposite the Foundling Hospital. It was laid out by S. P. Cockerell at the end of the 18th century when the Foundling estate was developed. The houses were put up by various speculative builders. A Victorian fountain, not without charm, stands on the axis of Guilford Place, but it lost much of its effect when a repulsive iron-railed convenience was constructed near by. The convenience, the statue and all the surrounding houses have been severely bombed, and the ruin of this part of Georgian London, which began with the demolition of the Foundling Hospital, is now practically complete.

BRIDGEWATER HOUSE, built in 1847, was the latest of Charles Barry's great Italian buildings in London. His biographer notes that it indicates " a desire for greater richness of effect " and the structure, though damaged, still stands to witness how Barry could express extreme affluence without becoming vulgar —an important asset in an architect much employed by Victorian aristocrats. The photograph shows the remains of the picture-gallery, designed to accommodate the famous collection which the last Duke of Bridgewater acquired from the Palais Royal after the Revolution and which passed to Barry's client, the Earl of Ellesmere. Although much of the interior of Bridgewater House was executed by a German architect, Götzenberg, on lines which Barry did not approve, the picture-gallery is authentic. Note the characteristic use of cast-iron principals for the roof.

Miscellaneous

ST. PAUL'S CHAPTER HOUSE, in St. Paul's Churchyard, was begun in 1712 from designs by Wren. Only the shell remains of this simple but extremely well-proportioned and delightful building. Its interior was little known to the public and had been shared by a bank and a club for many years. There was a noble staircase built round three sides of a rectangular well, with a rich wrought-iron balustrade executed by the smith, Thomas Robinson. Most of the rooms were panelled and the great room on the first floor had been redecorated in the 18th century. The top attic storey is, of course, a modern addition.

ESSEX STREET ARCH. Badly damaged, but now partly repaired, this forms the south end of Essex Street, Strand, one of the many speculations of Dr. Nicholas Barbon. It was not, as is so often stated, the " water-gate of Essex House," whose site was levelled by Barbon, but merely a closing feature to the street, probably designed by the Doctor himself, who included a working knowledge of the orders among less reputable accomplishments, mostly of a financial turn. Several of Barbon's houses, which were strictly standardised, even to the profiles of the balusters, survive in this pleasant street. Other houses of Barbon's, including those in Bedford Row, Red Lion Square, and chambers in the Temple, have suffered badly in the raids, and the visible memorials of this great figure in the history of the building industry are rapidly diminishing.

THE ADELPHI has lost some of the few surviving buildings of this great artistic and speculative venture of the Adam family. Adelphi Terrace, of course, was rebuilt before the war, and had previously been grotesquely mutilated by Victorian cement fiends. Mr. Stanley Hamp's massive office block took its place. But the buildings to north, east and west still stood. On the north, the Royal Society of Arts building remains, though slightly damaged. On the east there is No. 8 Adam Street, the lovely unspoiled façade, seen on the left of the picture, which ought to be preserved if it survives the war; while on the west there are the burnt buildings at the corner of John Street and Robert Street. These were not important architecturally, apart from some characteristic doorways.

ALBANY STREET was part of Nash's layout on the Crown's Marylebone estate. The illustration shows Nos. 152-4, built by Nash, in 1818, as a military Ophthalmic Hospital. It did not long serve this purpose, but became a laboratory for a primitive type of steam engine, a factory for no less primitive machine guns, and, later, though still in Nash's lifetime, a distillery. In recent years it had been let as workshops. On the left of the picture is James Pennethorne's ponderous but interesting Christ Church, Albany Street, dating from 1836-37. The tip of its spire has been knocked off.

Miscellaneous

THE RING, Blackfriars Road, had been famous for many years as a boxing centre, before its bombing and subsequent demolition. It was originally a chapel, built in 1783 for the Rev. Rowland Hill, a popular preacher of the time. It was a notable example of light timber construction within a brick shell, adapted for the purposes of a galleried auditorium. Rowland Hill was buried here, under his pulpit, in 1833, but when Christ Church, Blackfriars Road, was built his body was re-interred there, under the Lincoln Tower, and the old chapel was soon afterwards converted as a place of entertainment. The photograph provides a diagrammatic view of the building's structure.

BURLINGTON ARCADE, the northern part of which has been destroyed, was built in 1815-19, when Burlington House was remodelled for Lord George Cavendish. Lord George's architect, Samuel Ware, designed the Arcade, which is one of the few of its kind to have become and remained a commercial success. Both entrances of the Arcade had been rebuilt in recent years. The engraving shows the Piccadilly entrance before the rebuilding.

THE ST. KATHERINE'S DOCK was the last in the series of dock schemes which followed each other rapidly after 1800. The West India Dock was the first; then followed the London Dock, the Surrey Dock and the East India Dock. St. Katherine's was ambitiously squeezed in higher up the river than any of the others, displacing a mediæval college and a Georgian slum, just east of the Tower. The engineer was Telford, but Philip Hardwick was called in to design the Dock Offices shown in the illustration, which form part of the architectural scenery of Tower Hill and have been damaged by fire. They were completed in 1828.

WAREHOUSE ON TOWER HILL. Built to house some of the enormous wine and tea imports of the mid-Victorian age, this great warehouse took up its position opposite the Tower between 1865 and 1870. Mr. W. Potter, importer of wines and spirits, had his premises here before that date and lived next door. But at the rebuilding, William Potter, Esq., moved to a spacious house in Bayswater. The building always served a double purpose, part of it accommodating wines and spirits and another part tea. It was typical of what a number of architects—none of them famous—were turning out as the City people trekked out to the suburbs and offices and warehouses took over.

Miscellaneous

UNIVERSITY COLLEGE, GOWER STREET, built by William Wilkins, with Gandy-Deering, in 1827-28, has been partly burned. The photograph shows the rear of the main block, with the dome over the Flaxman Gallery, where a number of models by John Flaxman are preserved. The building to the right is the library designed by T. L. Donaldson, the College's first Professor of Architecture. The engraving shows the quadrangle with its famous portico, which stands intact.

"THE TIMES" BUILDING in Printing House Square, is said to have been designed by the second John Walter, chief proprietor of the paper and the one who employed Delane as editor. The erection of the building coincided with the formation of Queen Victoria Street in 1862, and the design, though amateurish enough, reflects very well the full-blooded Toryism of the day. There is something particularly attractive about the vigorous carved allegory in the pediment, in which a working clock takes its place with Father Time's scythe and a quantity of foliage as the insignia of "The Thunderer." The lettering incidentally anticipates the celebrated change-over to a Roman style heading made a few years ago.

PAGANI'S RESTAURANT, in Great Portland Street, burnt out in 1941 and now demolished, was a striking example of the late Victorian movement in favour of glazed terra-cotta, supported by such men as Halsey Ricardo and Beresford Pite. The first part of Pagani's to be built was the set of four arches on the ground floor, as a frontispiece to an existing building. This was in 1895, the architect being Charles H. Worley and the material Doulton's Carrara-ware. Five years later the premises were extended and Beresford Pite was employed to design a decorative façade for the whole, incorporating Worley's arches. The result was gay, appropriate and full of spirit. Other conspicuous buildings of the same period in similar material are the Birkbeck Bank (Knightley), Frascati's (Collcutt) and the Debenham house in Addison Road (Ricardo).

Miscellaneous

QUEEN'S HALL, Langham Place, was that rare thing, a successful concert hall. It was designed in 1887, begun in 1891 and opened with a smoking concert attended by Edward, Prince of Wales, in 1893. The plan was by C. J. Phipps, who designed so many London theatres ; the elevations were by T. E. Knightley. Knightley claimed the general conception of the hall, with its "trumpet-shaped" end and acoustically successful wall treatment, consisting of boarding separated from the wall and covered with composition over stretched canvas. Because of the light and air claims of All Souls' Church and schools the auditorium had to be partly sunk below street level, and this resulted in the pronounced squatness of the street front. The architectural treatment was highly ornamental in Mr. Knightley's inimitable and abandoned style. The interior, which was much modified in recent times to suit our less florid taste, was planned to accommodate an audience of 3,000, with 400 more in the orchestra. On the second floor was a smaller hall to hold 500.

BRISTOL
AND CLIFTON

Mary-le-Port Street, Bristol.

Bristol Churches

ST. MARY-LE-PORT CHURCH stood on a constricted site in Mary-le-Port Street (see picture on preceding page) and had but one aisle. Generous 15th century windows overlooked the small churchyard to the south, and there was a plain tower at the west end. The chief ornament of the church was a handsome wooden reredos, dating from the beginning of the 18th century, modelled, evidently, on the earlier one in St. Peter's (see below) but more refined in design and more sensitive in execution. Most of Bristol's old churches had these fine reredoses at one time, and the three best survivors, those at St. Peter's, St. Mary-le-Port and St. Nicholas, have disappeared during the raids.

ST. PETER'S CHURCH was typical of Bristol architecture at the height of the City's mediæval prosperity. Like the Temple Church and St. Mary-le-Port, it had no clerestory, but great rows of perpendicular windows in the aisles. The sturdy tower, which survives, is partly Romanesque, echoing the style of the once adjoining castle. The Corinthian reredos was an important feature. It was made by John Mitchell, a London joiner, who helped to furnish several of Wren's churches, and it set a fashion for such fittings in Bristol which persisted till the end of the 18th century. In the south aisle two elaborate tombs survive, though seriously damaged. The later of the two is that of Robert Aldworth, whose house, later known as St. Peter's Hospital, was burnt with the church (see page 84).

THE TEMPLE CHURCH
originally belonged, as its name
implies, to the Templars, but was
rebuilt after their time, in the
14th and 15th centuries. The
upper part of the tower (undamaged) was added about 1460
with alarming consequences,
the whole tower tilting sharply
westwards and coming to rest,
thanks to a rapidly improvised
internal buttress, 5 ft. out of the
perpendicular. The church's
chief treasure, a 15th century
candelabrum, has escaped.

ST. NICHOLAS CHURCH is a Gothic building of 1763-8
raised on a late mediæval crypt. Its large windows, with
simple tracery, are quite in the ancient Bristol tradition
and no doubt prompted Walpole's description of the church
as " neat and truly Gothic." The interior was also neat,
but even Walpole can hardly have found it truly Gothic,
for it was as perfectly Rococo as anything in England. The
ceiling, in particular, was a gay and skilful design, the work
of a famous west-country group of plasterers. The architect
of the church was Bridges, a local man. He incorporated
in the building the early 18th century reredos and some
wonderful wrought-iron-work from the old church. The
larger of the two pictures above shows the shell of the
church with its still standing tower hidden by ruined
buildings ; the smaller one shows the east end. In the
engraving, taken from Bristol Bridge, St. Nicholas is seen
in the centre with its " neat and truly Gothic " steeple. To
the left of it are Christ Church and All Saints (both undamaged). To the right is the tower of St. Mary-le-Port.

Bristol Domestic

THE DUTCH HOUSE was a conspicuous and admirable example of early 17th century domestic work, standing at the corner of Wine Street and High Street. It is traditionally supposed to have been constructed in Holland and shipped to Bristol, but the story is an unlikely one. The epithet "Dutch" was applied to any unwrought timber imported from the Continent and it is just conceivable, though hardly probable at the period, that the house was built with imported oak. The photograph shows the damaged façades hanging on the steel frame which had been introduced in recent years.

ST. PETER'S HOSPITAL, adjoining St. Peter's Church, was a medieval mansion, largely reconstructed in 1610 by the sugar-refiner, Robert Aldworth, whose splendid tomb still survives in the ruined church. In 1698 it was turned into a workhouse, one of the earliest institutions of its kind, and received the name of St. Peter's Hospital. As such it was used till 1865. It was later occupied by the Bristol Public Assistance Committee.

THE MERCHANT VENTURERS' ALMSHOUSES, in King Street, were built in 1696-8 at the expense of Bristol's great merchant-philanthropist, Edward Colston. They formed a small quadrangle adjoining the Merchant Venturers' Hall of 1701, which, together with part of the almshouses, has been distroyed. They are unpretentious buildings with a wooden eaves cornice and pretty carved door-hoods.

BERKELEY SQUARE was laid out in 1786, and built up slowly, houses still being offered for sale half finished in 1799. The Square is built on a gentle diagonal slope and the houses are stepped to accommodate the changing levels. Each separate cornice is carefully returned, and the string-courses stop against pilasters cleverly introduced to avoid the confusion of " loose ends " which would otherwise result. The pedimented feature in the centre of one side gives variety, though considering the one-way slope of the whole, such central emphasis seems hardly the thing.

Bristol

THE CASTLE of Bristol is represented by two vaulted rooms built into a modern shop in Tower Street and carefully preserved until the recent damage to the premises as a whole. Both these rooms probably belonged to the palace built in the castle precincts. This palace had a large Norman hall to which a porch—one of the two surviving rooms—was added in the 13th century. The other room, rather later in date and of plainer character, may have been the undercroft of the Royal Chapel.

FREEMASONS' HALL, Park Street, on the left in both the pictures alongside, was built as the Philosophical and Literary Institution in 1820-23. The architect, thirty-two-year-old C. R. Cockerell, gave the Institution the benefit of his Greek researches and attached a "tholos", with details suggested by what he had found at Bassae, to the corner of the building. The interior, burnt out, was of little architectural consequence. Beyond Cockerell's building are seen the jagged party-walls of houses in **PARK STREET,** one third of which has been destroyed. It was a fine, regular street, all of Bath stone, laid out in 1740 and built up from 1762 onwards. Sloping up north-west from the Green to the University tower (seen in the bottom picture) it could very well claim to be the handsomest shopping street in England.

THE UPPER ARCADE, together with the undamaged Lower Arcade, formed a street improvement carried out in 1824-5 and designed by the local Mr. James Foster, the architect of Clifton church (see next page). The line of these arcades linked up St. James's Barton and Broadmead. They were gracefully designed, with a pitched glass roof in compartments, but like so many of their kind they became rather drab commercial backwaters and never lived up to the distinction of their architecture.

THE MUSEUM owes its character largely, no doubt, to John Ruskin and *The Stones of Venice*. But it is not a fanatically Ruskinian building and its architects, Foster and Wood, derived from the Doge's Palace no qualities more profound than the decorative and the picturesque. Built in brown stone and yellow brick, the building has real charm. It was begun in 1866, a joint venture by two Bristol institutions, and only completed when taken over by the City in 1893.

Clifton Churches

CLIFTON PARISH CHURCH was built in 1819-22. It was paid for out of the sale of free-hold pew-rights, about two-thirds of the sittings in the church being disposed of in this way. The rich and largely non-resident population were prepared to pay handsomely for the prestige of a prominent and comfortable pew; the " free " sittings were largely occupied by powdered footmen. This curious financial adventure led to embarrassment when it was found that none of the pew-holders was willing to pay for the upkeep of the fabric. By 1884 it was decided that there was no alternative but to buy them out. This was gradually done and the church reseated. The architect of Clifton church was James Foster, a one-time assistant of the Bristol architect, Paty, and his design has all the characteristics of late Georgian Gothic. It may be compared with three other burnt churches of the same type, St. Nicholas', Bristol (page 83), and St. Luke's and St. Nicholas', Liverpool (pages 112 and 114). The two lower pictures show the staircases at the west end, where 18th and early 19th century monuments from the old parish church, built during the Commonwealth, were re-fixed.

ALL SAINTS, CLIFTON, was an important church by George Edmund Street, begun in 1863 and finished many years later. A curious feature is the abrupt junction of the nave arcades with the chancel aisles. This bold solution of the problem of combining a broad nave, suitable for Anglican worship, with a narrow chancel, may be compared with Street's very different treatment, in this respect, of St. John's, Kennington, (page 48). The materia's are local rubble, Bath stone and Pennant stone. The pulpit was by Pearson, 1892. The upper stage of the tower and a flamboyant lead-covered lantern were added by F. C. Eden in 1926-7.

Clifton Churches

HOLY TRINITY, HOTWELLS, was one of the least known, though not the least interesting, works of C. R. Cockerell. Built in 1829-32, when most churches were oblong boxes done up with Greek or Gothic detail, this one struck a new course. The plan was of the central-dome type so much used by Wren. The interior, somewhat altered by A. R. Gough, was pleasant enough; but the most strikingly beautiful part of the church was, and remains, the Italian entrance façade. Cockerell had the finest taste of any English architect of his time, and the delicate modelling of the niched entrance, the pilasters and pediment and the romantic little bell turret bear witness to this. A little earlier, Cockerell had designed the Hanover Chapel (long since demolished) in Regent Street, where he likewise adopted a Wren plan and developed it in his own charming and original way. If the Hotwells church is not rebuilt as a whole, the entrance front is a fragment well worth preserving for its beauty.

COVENTRY

The centre of Coventry, immediately after a heavy raid.

Coventry Churches

ST. MICHAEL'S, THE CATHEDRAL Church of Coventry since 1918, was wholly rebuilt in the 14th and 15th centuries at the expense of a single family of Coventry merchants, the Botoners. The magnificent tower, with its octagon and spire, was the first part to be built. It dates from 1373-94 and is the work of the Warwickshire school of masons whose " Perpendicular " was so accomplished and sensitive, and so unlike the routine products of the style which the Victorians very properly labelled " debased." Though St. Michael's had not the decorative qualities of the adjoining (unburnt) church of Holy Trinity, the tower of which appears in the small photograph on the right, it was impressively vast, with very broad nave and aisles, and additional aisles at the west-end, giving, from some points of view, a seemingly interminable perspective of arches and columns. It was a real merchants' church—huge, monotonous, generously windowed, with provision for many chantries and chapels for the merchant guilds. The destruction of the interior has been complete, even more devastating than the vigorous scraping and refurbishing which the building underwent at the hands of a local architect in 1851.

CHRIST CHURCH, apart from its beautiful octagon steeple, is a " Commissioner's Church," dating from 1829–32, and designed by Thomas Rickman, the pioneer exegetist of the Gothic styles, with his partner Hutchinson. The steeple is much older. Built by the Grey Friars in the 14th century, it was the only part of their buildings not demolished or quarried away after the Reformation. For centuries it stood alone in the fields outside the town until, when the land was threatened with building development, it was rescued to form the eastern termination of the new Christ Church. Rickman's work, though hard and thin, is in point of scholarship far in advance of other churches of the period. The safety of the tower has been in doubt, but it is hoped to secure its preservation.

Coventry

FORD'S HOSPITAL, one of the most famous timber buildings in the country, was built between 1509 and 1529, and originally housed single men and married couples—17 people in all. Latterly, it was occupied only by women. Few buildings of the kind have served their purpose so consistently and with so little alteration—and this in a great centre of modern industry. The architecture is remarkable, being the richest kind of domestic work done on a miniature scale. Though erected under the will of William Ford, the building of the hospital was the personal care of his executor, William Pisford, who considerably expanded its accommodation and endowment; and it is evidently to Pisford that the hospital owes its architectural elaboration. The hospital received a direct hit but escaped annihilation by fire, and can certainly be reconstructed, perhaps on a less inconvenient site. The pictures show the façade to Greyfriars Lane (top), the wrecked portion from the rear (centre) and the famous courtyard (bottom).

NO. 11, PRIORY ROW, a grand specimen of vernacular Palladianism of the second quarter of the 18th century, stands on the north side of the Cathedral, and belongs to the era when the dignitaries of Coventry were almost invariably mercers, drapers, silkmen or watchmakers. These were the trades which made the place prosperous in the 18th century and kept it prosperous till 1860 when its markets were devastated by the removal of tariffs, and Coventry, after a few years of acute distress, turned to bicycle manufacture. Priory Row makes an admirable "close" for the Cathedral and contains a number of fine houses, not all of which have been destroyed.

Coventry Domestic

NO. 16, LITTLE PARK STREET appears to have been built towards the beginning and dramatically "modernized" towards the end of the 18th century. The combination of coarse Palladian door and window dressings, with exaggerated "Adam" pilasters makes a highly unusual composition which would be found nowhere but in a centre of provincial wealth and provincial disregard for the rules of taste. The house illustrates the curious fact that the industrial towns of the 18th century far outstripped London in the externals of domestic architecture.

Though great engineering industries penetrated to the heart of the city, Coventry retained—and still retains— dozens of houses built for spinners, weavers and knitters in the hey-day of the English cloth-trade. Of these three wrecked houses, the two on the right (in Cox Street and Cook Street) appear to date from the early 17th century. That on the left (in Greyfriars Lane) belongs to the 18th century and is the type of house which would be occupied by an "undertaker" in the thriving silk ribbon industry of the period.

PORTSMOUTH

Portsmouth, with the burnt-out shell of St. Paul's Church.

Portsmouth Churches

THE GARRISON CHURCH is a mediæval building, and the loss of its modern roof is no serious matter, since the structure is otherwise practically unharmed. It formed the principal part of a "Domus Dei," such as still exists at Dover, at Chichester (St. Mary's Hall) and elsewhere. These buildings were hospitals and were mostly planned in the same way, with a nave-like structure in the aisles of which the beds of the sick were disposed, and, at the east end, a chapel. The chapel at Portsmouth, which makes the chancel of the Garrison Church, has an excellent 13th century vault, quite unharmed by the fire. The "nave" was the hospital proper. The resemblance of the whole plan to that of an ordinary parish church is an interesting comment on the limited building-formulas of the Middle Ages. The church contained a large number of military monuments, some of which have survived.

ST. GEORGE'S, PORTSEA, hopelessly wrecked by blast, was built for their own use by a group of shipwrights in 1753. Its architecture is quaint and unorthodox, suggesting that they made their own design. Its galleried interior had four great columns supporting the roof, and retained much of its 18th century atmosphere.

ST. PAUL'S stands in St. Paul's Square, Southsea (see next page). It was one of the many churches built at State expense under the Act of 1818, and dates from 1820-22. The designer was Francis Goodwin, whose most famous work was the old Town Hall at Manchester, a clever Greek building. Many Gothic churches of his are to be found in the Midlands, where he was chiefly employed. He was fond of cast-iron, whose rigid lines and sharp arrises suited his conception of Gothic very well. Here at Southsea, all the tracery is in this material, founded locally. There were, of course, galleries, and the lines of these and the ribbed plaster vault can be seen on the east wall. The masonry was in Bath stone.

Portsmouth Domestic

THE HIGH STREET of Portsmouth, with its many Georgian houses and general air of belonging to the age of Trafalgar, has been largely destroyed. The picture above shows a section of the street, with the Museum and Art Gallery, which was originally the Guildhall. It was begun in 1837 and opened on Queen Victoria's coronation day; and though the taste of the architect, Benjamin Bramble, was neither up-to-date nor correct, the little portico made a pleasant feature in the street as it was.

ST. PAUL'S SQUARE, Southsea, was a pleasant stucco-fronted square of the early 19th century, with some pretty Gothic houses on the west side and a hall on the east. The railings of the church in the centre (see facing page) appear on the left of the photograph.

Portsmouth

THE UNITARIAN CHAPEL in the High Street was built in 1717–18 but has a history going back to the early days of non-conformity. Unlike most chapels of its date, it occupies a conspicuous site and its congregation has always played an important part in Portsmouth history. Much of the timber in the interior came from the building's 17th-century predecessor. Like many of the older chapels, it passed from being Presbyterian to Arian and thence Unitarian under successive ministrations during the 18th and early 19th centuries, when its alteration was recorded with a tablet dated 1822.

THE GUILDHALL is a big classical block, belonging, architecturally, to the group of Victorian town halls in which those at Leeds, Halifax and Bolton are the most conspicuous. Built in 1886–90, it is the latest of them, but it adheres to the old fashion, and reflects hardly a single nuance of late Victorian taste. Its architect, William Hill, came from Leeds and evidently derived much from Brodrick's Town Hall in that city. But the Portsmouth building lacks serenity, and Brodrick's strong, vigorous detail. The interiors, all completely burnt, were heavy and dull, and the building as a whole instances the deadness of a design wholly detached from the influences of its time.

PLYMOUTH

Household wreckage at Plymouth.

Plymouth Churches

ST. ANDREW'S CHURCH stands in the centre of Plymouth, its roof and furnishings burnt and its monuments badly damaged. It is a grand early 15th century building, with aisles running the whole length and short additional aisles on north and south giving a transeptal effect on plan. The dominating west tower was built at the expense of a Plymouth merchant in 1440–60. Granite has been much used, particularly in the beautiful arcade capitals and the dressed masonry of the tower. The fine effect of the arcades and continuous barrel roof is well seen in the engraving, taken shortly after Foulston's remodelling of 1826. The church was taken in hand by Sir Gilbert Scott in 1874–5 and Foulston's work, with other Georgian features, removed. The monuments in St. Andrew's are of great interest and range from elaborate 16th and 17th century mural monuments to the bust (slightly damaged) of Dr. Zachariah Mudge, executed by Chantrey from a painting by Reynolds. Adjoining the church is the mediæval Prysten House, which has been badly damaged though not burnt.

CHARLES CHURCH, which is in much the same state as St. Andrew's, was built under an Act of 1640, creating a second parish in Plymouth and stipulating that its church should bear the King's name. The body of the church was finished in 1657 and the tower in 1708, with a wood spire, replaced by a stone one in 1766. The nave and aisles, modelled, no doubt, on St. Andrew's, are as correctly Gothic as anything one could find in the 17th century, and the geometrical east window is remarkably good. Other windows were put in in 1864 when a rather heavy restoration was undertaken. The monuments, which have been severely damaged, were of some interest.

NORLEY CHAPEL was built in 1797 as the New Tabernacle by a Methodist congregation of the Calvinistic persuasion whose inspiration came from Whitefield rather than Wesley, and was an offshoot of the Old Tabernacle whose shell still stands in Exeter Street. It was later covered with stucco.

THE UNITARIAN CHURCH, Treville Street, has a history going back to 1662 and the Act of Uniformity, when Hughes, the leader of Devonshire puritanism, founded his congregation. A section of this congregation adopted Arian and subsequently Unitarian views early in the 18th century and the Treville Street church became theirs exclusively. It was rebuilt in its present form in 1832. The spire of the burnt-out Charles church (see above) appears in the background of the photograph.

Plymouth

GEORGE STREET BAPTIST CHAPEL (left in the illustration) stands in the ruined centre of Plymouth. It was built in 1845 for a congregation whose history goes back to 1620-40 and which is the oldest of the many old nonconformist congregations of Plymouth. This photograph was taken from the tower of the Guildhall.

THE GUILDHALL, burnt and blasted through its whole length and breadth, consists of two blocks of hard, robust Victorian Gothic. The north block, illustrated here, contained the Council Chamber and Municipal Offices. The south block contained the Guildhall proper and Assize Courts. Between the two is a public space. The building was the result of a competition, held in 1869, with Alfred Waterhouse as assessor. The winners, out of twenty competitors, were the Plymouth architects, Norman and Hine; but the artistic responsibility for the building was placed in the hands of E. W. Godwin, whose town halls at Northampton and Congleton were thought well of by the critics. Godwin was a man of fine taste, a devotee, like his friend Burges, of " early French," and unlike most of the revivalists, more interested in secular than church work. There is some good detailing and well placed sculpture in the Plymouth building, though as a composition it is diffuse and dull—like most things of its date and kind. The materials are mainly local limestone and Cornish granite, with Portland and Mansfield for the carved work —a typically Victorian undervaluation of the potentialities of granite.

THE OLD GUILDHALL was built in 1800 by Eveleigh, a Bath architect, who optimistically undertook to provide, on a small and hideously inconvenient site, a building comprising court-house, guildhall, mayoralty house, civil and criminal prisons, guard rooms, news-room and market. The result was a dreadful failure from every point of view and Eveleigh, who did creditable work in his own city, earned nothing but discredit from the people of Plymouth.

Plymouth

THE ROYAL HOTEL formed part of a great island block of buildings comprising, besides the Hotel, the Assembly Rooms and Theatre. It was the work of John Foulston, Plymouth's great architect, and was projected and paid for by the Corporation. It was begun in 1811 and finished in 1813, when the Theatre opened with *As You Like It*. The block, with its two grand Ionic porticos, and the contrasting Doric of Foulston's Athenaeum (also burnt) alongside, was a fine thing when it was complete. But, before the war, Plymouth had chosen to destroy the Theatre and allow a cinema to be built which negatived all possibility of future unity and order. The burning of the remainder of the block is, therefore, hardly an architectural disaster. The interior of the Hotel was of no distinction, but the Assembly Rooms had rich and beautiful Regency details.

LOCKYER STREET, designed by Foulston and opened in 1821, perpetuates the name of the family under whom, as Mayors, so many Plymouth improvements were set on foot during the Regency. Like a number of Plymouth's neat stucco streets it has been gutted almost from end to end.

PRINCESS SQUARE is another piece of Foulston's work to have been almost completely destroyed. It was part of the lay-out which developed between the Hotel and Theatre block (see above) and the Hoe. These Plymouth improvements are so closely contemporary with Nash's improvements in London as certainly to represent a quite separate initiative.

MANCHESTER

The Market Place, Manchester, after a raid, showing the old Wellington Inn and, in the distance, part of the Cathedral.

Manchester

A demolished island site in Manchester. Here stood the **VICTORIA HOTEL** and **VICTORIA BUILDINGS**, typical grotesque Manchester Gothic of about 1872, when Deansgate, the street seen on the left, was widened and Victoria Street, on the right, formed. The Cathedral tower, seen in the distance, is a rebuilding of 1866 on approximately original lines.

THE CATHEDRAL received a direct hit on the Derby Chapel, which adjoins the Lady Chapel, from which the photograph was taken. This fine church, Manchester's Cathedral since 1847, dates from the first half of the 15th century, when the old parish church was handsomely rebuilt on collegiate lines, with long nave and aisles continuing with chancel and chancel aisles. It is particularly notable for its wonderful woodwork, including pulpitum, screens, parcloses and misereres, mostly undamaged and now protected. The woodwork all belongs to the period 1485–1506 and is marked by great decorative skill and refinement. The monument seen in the photograph is that of Humphrey Chetham, erected in 1853, the bicentenary of his death, the sculptor being William Theed. It was Chetham who, by his will, enabled the buildings of the mediæval college to be converted into the school still known as Chetham's Hospital (see page 110).

FREE TRADE HALL, one of Manchester's finest classical buildings and one of the noblest public halls in the country, was among the last works of Edward Walters, who was responsible for so many of the great Italian warehouses in the city. Completed in 1856, it occupies the site, and inherits the name, of a timber hall built by the Anti-corn-law League for meetings in support of the Free Trade movement. The great hall, whose burnt interior is seen above, had a flat ceiling which met the wall above the arches in a deep cove. The exterior to Peter Street was, and is still, a wonderfully fine Italian composition whose arched treatment is a variation of that of the hall itself. The building contained a smaller Assembly Hall at first floor level.

THE ROYAL EXCHANGE, the successor of two earlier buildings serving the same purpose, was completed by Mills and Murgatroyd in 1874 but entirely remodelled in 1914-21 by Bradshaw Gass and Hope, when the Corinthian portico was removed and the site extended to the south. The main hall, seen here, accommodated 12,000 to 13,000 people and was among the largest covered spaces of its kind in the country. The detail is a revised version of that of 1874 and the old clocktower, also seen in the photograph, was retained.

Manchester

CHETHAM'S HOSPITAL has been slightly damaged by incendiary bombs. Historically by far the most interesting of Manchester's buildings, it was originally the priests' college connected with the Collegiate Church, now the Cathedral (see page 108). At the Dissolution it became the property of the Derbys, but was acquired in 1653 for the use of Chetham's foundation. The older parts, in heavy Lancastrian Gothic of the 15th century, include the hall and its porch seen in the photograph.

CROSS STREET CHAPEL was Manchester's oldest and principal dissenting church. Its red brick shell and some of the fittings survive. Within living memory, it was the most fashionable place of worship in the city, attended by the richest cotton and banking families. Built in 1694 for Henry Newcome, a preacher ejected from the Collegiate Church at the Restoration, it was partly destroyed in a riot in 1715, but rebuilt at Government expense. At this date the fine pulpit, seen in the photograph, was introduced. It is almost identical with a pulpit of the same date in St. Anne's church, nearby. Like other early dissenting chapels, Cross Street Chapel was at first Presbyterian and in latter days Unitarian, this being the denomination most favoured by the professional and upper classes in the early 19th century. It is one of many historic nonconformist buildings which have suffered in air-raids, including examples at Plymouth, Southampton, Portsmouth, Hull and London.

LIVERPOOL

The ruins of Burlington House, Liverpool, still smouldering after a raid.

Liverpool Churches

ST. NICHOLAS' CHURCH stands by the water's edge near the docks, and its beautiful west tower is still a landmark in spite of the bombing and in spite of the competing hugeness of the buildings along the front. This tower, at once masculine and decorative, with an admirably proportioned lantern, is the work of Thomas Harrison of Chester. It succeeded an ancient steeple which collapsed at service time one Sunday in 1810, killing a number of charity children but sparing more influential and providentially less punctual members of the congregation. The body of the church is older; it was built under the direction of Joshua Brooker in 1774-5, on the site of a very plain mediæval building which belonged to the days when Liverpool was a mere chapelry of Walton-on-the-Hill. The interior of St. Nicholas' was of no great artistic account, but the monuments were interesting and these have been salvaged. The illustration on the right shows the interior looking towards the tower, with the slim cylindrical columns which may be compared with those in Clifton Parish Church (pages 88-9).

ST. MICHAEL'S, PITT STREET, which has now been wholly demolished, stood in a square in the area of Liverpool developed between 1790 and 1810. The church itself was built in 1816-26, by the great Liverpool architect, John Foster, in partnership with his father. The steeple, with its four " portals " embracing the base of a tapering spire, was a thoughtful and interesting design, an attempt to purify the tricky generalisation of the Gibbs model on which it is so evidently based. As a ruin, this church, with its magnificent masonry and Corinthian portico, was very wonderful. In the illustration, Liverpool Cathedral can be seen in the distance.

ST. CATHERINE'S, whose Greek Ionic portico occupies the centre of one side of Abercrombie Square, was another work by John Foster, who added so many beauties to his native city. It was begun in 1831 by a company largely formed from the rich householders of the Square. The interior has been burnt out, but the portico survives.

Liverpool Churches

ST. LUKE'S CHURCH stands conspicuously at the head of Bold Street, for whose well-to-do inhabitants it was built. This church and St. Michael's (previous page) were originally projected as a pair in 1793 and built many years later at the joint expense of the parishes and the Corporation. The direct part which the Corporation and its architect, Foster, played in the public works of this period, is remarkable. Foster designed St. Luke's in 1802 and began it in 1810, but its west tower (see the engraving below), was not finished till 1831. It is a rather harsh Gothic building and the tower, in spite of much enrichment, is graceless and forbidding. The interior was notable for its timber columns and ingenious plasterwork.

THE CATHEDRAL, right, designed by Sir Giles Gilbert Scott, and begun in 1904, is so vast a fabric that the injuries it has received are scarcely noticeable, though, as the illustration of damage in the Derby Memorial Transept shows, they are not negligible. To appraise the extraordinary qualities of this building, a meteor from the eclipsed firmament of Gothic revivalism, will never be easy, and is to-day impossible. There is, perhaps, no building in the world which reflects, on a comparable scale, the changing mentality of a generation focussed at the drawing-board of one man.

THE CUSTOM HOUSE, designed by John Foster in 1826 and begun two years later, is a cold, hard, but extraordinarily fine building. It occupies the site of an old dock and was built at the joint expense of the Corporation and the Government. Conceived in a lordly way, without parsimony or restriction, it comprised the excise, stamp and dock offices of the port. Its centre and two great wings cover a vast amount of ground, but the group is well-knit and the rotunda, which might so easily have been a pretentious irrelevance, sits majestically over it. A comparison with Smirke, the leading exponent of unadorned Ionic, is inevitable; but there is a gravity in Foster's work where Smirke is so often merely dull.

BLUE COAT CHAMBERS once housed Captain Bryan Blundell's charity school, founded in 1708 and established as the Blue Coat Hospital in 1718 when the school building was begun. Sixty children were provided with food and clothing, instructed in the three Rs and, to ensure their usefulness to the community, in cotton-spinning. The building, in red brick with stone dressings, has the virtues of its vernacular kind and the fenestration of the wings is especially good. After the last war the building was threatened with demolition, but rescued and let as offices, and the Liverpool Architectural Society had its headquarters here up to the present war.

Liverpool

ST. GEORGE'S CRESCENT recalls once again the architectural dictatorship of John Foster. He replanned this part of Liverpool in 1825-7 and formed this crescent at the foot of Lord Street, providing a good setting for the classical church of St. George which stood on the site of the present Queen Victoria memorial. The buildings were stucco-fronted, with an Ionic order embracing the upper storeys in pilaster form.

THE GOREE WAREHOUSES, on the waterfront of George's Dock, are among the finest examples of early 19th century industrial architecture. Though altered from time to time and lacking, as a result, the impressive uniformity evident in the engraving, they retain their most striking feature, the "piazzas" with their alternate large and small arches. This traditional English misuse of the Italian word derived from Covent Garden, where it became attached to Inigo Jones's arcades instead of to the square itself. The grooves in the stones, perhaps designed to facilitate lifting, give an unusual textural effect to the masonry. The warehouses were begun in 1802 after a disastrous fire which demolished an older and even taller range of warehouses.

OTHER LARGE TOWNS

A burning chapel in Belfast.

Birmingham

THE CATHEDRAL of St. Philip was built in 1711-19 as the church of a new parish. The architect was Thomas Archer and this is the earliest of his three great Baroque churches. The second was St. John's, Smith Square, London (see page 36) and the last, St. Paul's, Deptford. St. Philip's is an aisled church (unlike the other two, which are on the Greek cross principle), but the whole conception is uncompromisingly classical, except for the tower, and this follows tradition only to the extent of being at the west end. The giant Corinthian pilasters give a hint of the kind of interior which Archer evolved for his later churches; some of the details are Baroque of an extraordinarily advanced and complicated character and suggest that Archer had studied 17th century Roman work more closely than any architect of his time. St. Philip's, which became a Cathedral in 1905, is famous for its series of windows designed by Sir Edward Burne-Jones, fortunately removed before the raids. The roof was badly damaged by fire but has been temporarily repaired.

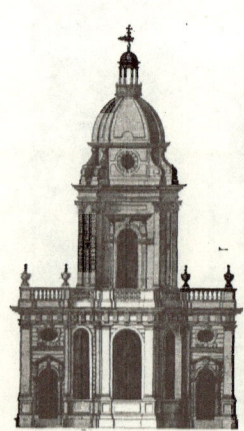

ST. THOMAS'S CHURCH, Edgbaston, was a Commissioner's church, built in 1825-7 by Thomas Rickman with his partner and former pupil, Hutchinson. The west front, still standing, is an unusual composition, consisting of two quadrant porticos flanking the main entrance in the base of the tower, an arrangement which ingeniously avoids the difficult relationship of tower over portico, without sacrificing the portico idea. Classical works by Rickman are rare, for he was a learned Goth and it is, of course, to him that we owe the popular terminology of Gothic architecture—" Norman," " E.E.," " Dec." and " Perp.": terms which, inaccurate as they may be, have been in use for 130 years and are not likely to drop out just yet. A typical Gothic church by Rickman is illustrated on page 93.

THE MARKET HALL was designed by the Birmingham architect, Charles Edge, and opened in 1835. It was one of many classical markets erected in the first half of the 19th century. The earliest of them was John Foster's still existing market at Liverpool (1819-20), the finest those by Fowler at Exeter and Covent Garden. The Birmingham market had three " aisles " separated by rows of cast-iron columns, survivors of which are seen in the picture on the right. It provided accommodation for 600 stalls—and was lit by gas.

Southampton

HOLY ROOD CHURCH, standing conspicuously in Southampton's main street, was practically rebuilt in 1848-50 by the architects Mee and Webb. The tower and spire and some of the foundations and walls of the mediæval church were retained, but the structure as it stood before the bombing was of very little interest. As a result of the bombing, a few unsuspected mediæval features have been revealed.

ALL SAINTS was a very remarkable 18th century church, on the site of a mediæval predecessor. Its architect, almost unknown to fame, was Willey Reveley. He was a pupil of Sir William Chambers, but travelled widely enough to revise some of his master's prejudices. He began All Saints in 1792, when church-building was at a low ebb and the Wren-Gibbs tradition was wearing very thin. Instead of the usual aisled arrangement, he constructed a segmental, coffered ceiling covering the entire 61 ft. span of the church, and the chancel was made to occupy the space under an eastern cupola, the weight of the latter being adroitly distributed in the manner shown in the illustration. The church was gutted and is to be completely demolished.

ST. MARY'S is the imposing modern successor of two older churches, the first one, mediæval, destroyed because it served as a landmark to the French enemy from the sea, and the second a Georgian brick box. Gothicised in the 19th century and finally replaced by the present structure in 1878-9. The architect was George Edmund Street, and, like most of Street's later churches, St. Mary's is very English, very correct, beautifully detailed and rather tame. The work was carried on by the architect's son, A. E. Street, and completed in 1884 with the extremely handsome western tower and spire which still survive. The body of the church has been gutted and, as the illustration shows, the masonry has been badly damaged. St. Mary's was built as a memorial to Bishop Wilberforce of Winchester.

Nottingham

ST. JOHN'S, Lean Side, was an "Early English" church designed by George Gilbert Scott with his partner Moffat and built in 1843-4. It thus belongs to Scott's early days, when he was struggling with the "true principles" and trying to forget the poor-law institutions and spiky suburban churches with which he began his amazing career. St. John's represented a substantial advance, but was still not quite the thing: the over-emphasised central lancet and trivial rose-window are witnesses of immaturity. It is instructive to compare Scott's east end here with Street's at St. Mary's, Southampton (previous page), where the same theme is handled, after an interval of more than thirty years, with perfect assurance and perhaps more skill than Scott commanded, even in his later years.

UNIVERSITY COLLEGE, part of which has been destroyed, was built in 1877-81 to comprise a public library, museum and lecture theatres, the whole forming part of a scheme for the education of the working classes. The architects were W. and R. Mawson of Bradford, and the style they adopted rather resembles the Anglo-Venetian blend of Gothic invented by Sir Gilbert Scott. The College is a successful building in its high Victorian way, and the masonry, in warm Ancaster stone, is admirable.

Hull

THE CHARTERHOUSE is a building in the excellent taste of 1780, with a chapel in the centre and lodgings for the brothers and sisters on either side. The foundation dates from the 14th century when the hospital was established outside the Carthusian monastery from which it takes its name. The present building has suffered badly from blast but is not beyond repair. The illustration shows the porch to the chapel. Above the pediment is a well-designed cupola; there is a remarkably fine pulpit inside and the rooms retain their original combination grates for heating and cooking.

TRINITY ALMHOUSES, designed by the Hull architect, Charles Mountain, in 1834, were scheduled for demolition before the war. This was one of several fine buildings for which the Trinity House of Hull was responsible in the 18th and 19th centuries. It comprised thirty-six apartments for decayed mariners and wives and widows of mariners. The stucco façade was considerably higher than the structure warranted and absurdly out of keeping with the bare simplicity of the rooms within. Nevertheless, it supported the dignity of Trinity House and was a striking adornment of the not very beautiful city of Hull. Excellent and more appropriately designed brick almshouses, built a few years later, survive along two sides of the site of the main building, which has been entirely destroyed.

Hull

THE CONGREGATIONAL CHAPEL stands in Albion Street and has been gutted. It was the work of H. F. Lockwood, 1842, the architect who, with the Mawson brothers, designed the City Temple, London (see pages 50-51) thirty years later. The Chapel was denominated Independent on its foundation, but eventually became Congregationalist in character.

THE PILOT OFFICE was built in red brick shortly after 1832, on the quay overlooking the Ferry Pier. Administered by commissioners under the Humber Pilot Act, it was originally staffed by forty-six pilots, who took duty in rotation in the observatory on the top floor. The building has been badly damaged by blast and the decorative masonry has fallen from the window over the columned entrance.

GEORGE STREET, Hull's principal Georgian street, was part of a spacious development of the town, dating from about 1790, and comprising also Kingston Square and Jarratt Street. The houses are of the dull red brick common in the north, and most of them have porches. On the north side, the monotony of the street was broken by the series of large houses seen in the illustrations (left and centre) with pediments filled with medallions and swags. The interiors had spacious rooms almost entirely without ornament. On the right is a view of part of Little Humber Street, the south-westward continuation of Hull's ancient High Street. This building incorporates the Little Lane archway, belonging to the 17th century or earlier and originally forming a water-gate to the Humber.

WAREHOUSE behind Wilberforce House Museum, in High Street. This had been transformed into an extension to the Museum and contained a street of old house fronts transferred from different parts of the country and including, for example, an exceptionally fine Georgian shop front from Lewes, Sussex. Its chief interest, as seen in the illustration, is that it was typical of the brick and cast-iron warehouses built in the 19th century in the gardens of the old merchants' houses between High Street and the Humber. Wilberforce House itself, dating from Charles I's time and containing excellent wood and plaster work typical of 18th century Hull, has luckily survived intact. The warehouse was designed by William Botterill, architect of the Hull Exchange, in 1865.

Croydon

ST. ANSELM'S SCHOOL, Croydon, was an early Georgian house, sufficiently distinguished in all its parts to be considered of more than average merit. The five-windowed front, with rich cornices and a carved door case, was beautifully proportioned and detailed; the wrought-iron entrance gate, whose remains are seen on the right of the picture, was correspondingly good, and so were the staircase and panelling of the interior. The house is supposed to have been built in 1708 and was at one time associated with a Friends' school.

Swansea

THE WESLEY CHAPEL is a curious witness of stray influences which have wandered into Wales from early 19th century England. The high attics, bracket cornices and rustications come from Barry's Italian revival, but the porch at the end harks back further, to Sir John Soane. The assertive gable thrown up over the main entrance is a native gesture. Wales had no tradition of professional architecture before the 19th century, and the study of a building such as this, if the facts attending its erection were known, would provide an interesting psychological footnote to Welsh history.

THE MARKET, designed in 1889, was a late Victorian equivalent of the type of market built at Birmingham (see page 119) in 1835. It consisted of an enormous covered space within an ornate Renaissance shell, and was planned with a central meat and flower block round which was a continuous space for stalls, roofed with light iron and steel trusses of about 60 ft. span. The design was selected in competition, the architects being J. Buckley Wilson and Glendinning Moxham, of Swansea.

East Ham

ST. MARY'S CHURCH, though severely damaged, remains one of the chief archaeological treasures of Essex. It is a complete 12th-century building, with an apse; the tower was built in the early 16th century. The interior is full of beautiful details and fittings, including an interlacing Norman wall-arcade, 13th-century paintings and 17th-century monuments.

COUNTRY
AND SMALLER TOWNS

A lodge-keeper's cottage in the West of England.

Eton College

The most serious destruction at **ETON COLLEGE** was caused by a direct hit on the Saville House (facing page), a building erected in 1603-4 and intended to contain Dr. Saville's printing presses. The picture shows the Saville House across Weston's Yard, with " Weston's ", a Tudor building much altered, beyond and the dome of the South African War Memorial building on the left. Damage was also sustained by the Upper School (photograph below), a building of 1689-91, containing much of the original interior, including the old headmaster's classroom whose windows are seen on the first floor in the photograph. This is Eton's principal renaissance building and forms the west side of the School Yard; the ground floor is partly open, with a colonnade opening on to the Yard.

Canterbury

Bombs fell dangerously near Canterbury Cathedral, but only shattered some relatively modern glass of little importance. The buildings in the precincts and neighbouring streets, however, sustained damage. The view on the left shows where some old houses in **BURGATE STREET** were demolished. Beyond the ruins are the backs of houses in the Cathedral Churchyard, while in the distance is the Cathedral itself. In the Green Court there is more damage. **THE DEANERY** (right-hand picture), a building of Prior Goldstone's time (1495-1517), containing good 17th century work, has an ugly gap in its façade, and there was also damage at Chillenden's Lodging in the south-west corner of the Green Court.

Yarmouth

GREYFRIARS consists of two bays of the Friars' cloister, with some slight remains above, all dating from the 13th century and hidden from public view until bombing opened them to the daylight. On the ground floor is a complete vault with wall-shafts and moulded capitals and ribs. Some of the upper part is seen in the illustration.

THE TOLLHOUSE is a secular mediæval building of special interest because of its association with municipal government at least as far back as 1362. Parts of the building, however, are more than a century earlier than that. There is a hall, approached by an external staircase, substituted, presumably for a similar feature, in 1622. Below the hall is a prison with a separate entrance. A 17th-century addition formerly existed over the passage-way on the right. The building was restored in 1885 and became a museum shortly afterwards.

Newmarket

HIGH STREET, Newmarket, described in 1692 as "full of Inns," and still full of them, is a typical product of the kind of town which developed along a great highway in the 17th and 18th centuries. Before the coming of the railways, the White Hart, seen in the illustration, was the great hostelry of Newmarket. It stood on the north side of the street, to which it presented a stucco front, with an archway in the centre leading through to the stable yard.

Derbyshire

WOODSEATS HALL, a lonely Derbyshire farmhouse among the hills near Barlow, north-west of Chesterfield, has been seriously damaged. It is a typical Derbyshire hall, built of local grey limestone in the early 17th century. Among the adjoining buildings is a fine old barn with a roof supported on crucks.

Dover

ST. JAMES-THE-LESS is a church of Norman origin, with some early features still remaining. It was abandoned during the last century, when a florid modern church of St. James was erected; the old building, however, was vigorously restored and put once more into use. The damaged transept, seen in the illustration, is almost entirely modern work, and the church as a whole cannot be accounted of great architectural importance.

THE ROUND HOUSE is a pretty Regency villa standing between long rows of sea-side terraces and looking out across the channel, an intriguing incident in the panorama of Dover as seen from a channel steamer. It was built in the early 19th century for Mr. John Shipdem, who was chosen Town Clerk of Dover in 1791 and held the post, together with that of Register of Dover Harbour, for thirty-five years. The architect is unknown, but if it was not John Nash it was somebody who followed Nash's ideas very closely. The Round House bears a marked resemblance to two of his houses, particularly to the Casina he built for Richard Shaw of Dulwich in 1797-8. It had long been abandoned as a residence and, before the war, was in use as a Gospel Hall.

Kent

STURRY is a little village on the Stour, not far from Canterbury, with old houses and inns tightly packed along its narrow street. None of them is, or was, of architectural importance, though as a group Sturry typifies the beauty of Kentish vernacular architecture with its high average standard of design and craftsmanship.

Brighton

NORFOLK SQUARE joins one of the many Brighton streets which go up from the sea and are filled with tall, narrow houses with bow fronts, designed so that a glimpse of the channel might just conceivably be caught from one window on each floor. The part of Brighton where Norfolk Square lies was laid out about 1820-30 on no particular plan other than the obvious one of providing as many houses as possible within the shortest possible distance from the sea. When these houses were built, the glories of Regency Brighton were already fading, the King was rarely at his Pavilion and the mass invasion of the merely rich had begun.

Country Churches

BROMLEY, KENT. The oldest portion of the parish church of SS. Peter and Paul was the fifteenth-century tower, which is the one portion that has escaped total destruction. The tower is of flint, but the body of the church had been remodelled in red brick in 1792. It was again largely rebuilt in 1830 and restored in 1873 and 1884, with the result that the Georgian rebuilding, though good of its kind, had lost its characteristic simplicity. The tower, which houses a peal of eight bells hung in 1727, was itself set alight but, as the picture shows, damage to its structure was not great.

SANDERSTEAD, SURREY. All Saints is an Early English church with later additions including tower and spire in typical Surrey style. Built of flint and stone, it was considerably restored in 1846. During enlargement in 1938 some early 14th-century mural paintings were discovered. Structural damage has been chiefly confined to the roof.

GREAT COGGESHALL, ESSEX. The church of St. Peter-ad-Vincula is a very fine example of the East Anglian Perpendicular style, with an embattled flint and stone tower with an angle turret. It is very consistent in style, the only considerable later addition being an ornate stone and alabaster reredos put up in 1880. A direct hit did extensive damage to the west end of the church. The large picture is taken from the chancel; the small one is an exterior from the north-west. Great Coggeshall contains the famous Paycock's house, a remarkable example of decorated timber architecture. A brass effigy of Thomas Paycock, dated 1586, is in the church.

Country Churches

PAKEFIELD, SUFFOLK (adjoining Lowestoft, on a stretch of coast which has lately been considerably eroded). This church has an unusual plan, being divided longitudinally into two equal-sized naves, which were originally independent churches, All Saints' and St. Margaret's. An elevated chancel has a crypt beneath. The church was of flint with a thatched roof and has been almost destroyed. The thatch can be seen still burning in the pictures above.

DIBDEN, HAMPSHIRE, a parish near Southampton Water, has a tiny isolated church with a farm nearby. The church, which has been burnt out, has a 13th-century chancel, arcades and aisles of later date and a tower built in 1884. The picture, taken from within the damaged nave, shows the tower arch, with the modern west window beyond.

STOW-BEDON, NORFOLK (a village in the southern part of the county). St. Botolph's church is a considerably restored Perpendicular building with two early Decorated windows and an Early English east window. The western porch seen in the picture is an addition of 1887. The best features of the interior are a carved oak rood screen and some stained glass removed from Hildersham, Cambridge. Bomb damage is chiefly confined to the roof.

136

SOUTHWICK, SUSSEX. Serious damage to the parish church of St. Michael and All Angels is confined to the very fine Norman and Transitional west tower shown in the picture. The walls of the tower were badly cracked from top to bottom by a bomb which fell in the churchyard between the west end of the church and the rectory, so much so that the upper portion is being removed, the north-west part of the tower having sunk several inches. The rest of the church is of less interest than the tower, having suffered considerably from 19th-century restoration, but the interior contains several features of value including a Jacobean pulpit.

Llandaff Cathedral

LLANDAFF CATHEDRAL, on the outskirts of Cardiff, is unusual for having no central tower and no transepts, so that nave and choir are all under one roof. The nave is Early English; the choir is Decorated; so is the lady chapel. There is no triforium. The best features are the graceful Early English west front, the late Norman doorways to north and south of the nave and, in the interior, the late Norman arch at the back of the choir. There is an unusual Chapter House : square in plan, with quadripartite roof and a central column. The south tower collapsed in a storm in 1703 and was rebuilt in 1869 by the Welsh architect, John Prichard. The whole cathedral had been allowed to fall into a bad state of repair following the Reformation, and in 1732 John Wood of Bath built a classical temple within its walls. In an air raid on Cardiff, the whole south side of the nave was unroofed, the interior considerably damaged, nearly all the windows, including the large west window, blown out and the roof of the Chapter House, on the right-hand side of the picture, destroyed.

INDEX

Adam Brothers, archts., 75
Archer, Thomas, archt., 36, 118

Bacon, sculptor, 6
Baily, sculptor, 12
Banks, sculptor, 18
Barbon, Dr. Nicholas, 59, 74
Barlow, Derbyshire—
  Woodseats Hall, 131
Barry, Sir Charles, archt., 52, 73
Bedford, Francis Octavius, archt., 42
Bernasconi, sculptor, 71
Birmingham—
  Churches:
    St. Philip's (the Cathedral), 118
    St. Thomas's, Edgbaston, 119
  Market Hall, 119
Blomfield, Sir Arthur, archt., 30
  Sir Reginald, archt., 29, 71
Blore, Edward, archt., 11, 63
Bodley and Garner, archts., 12, 46, 48
Bradshaw Gass and Hope, archts., 109
Bramble, Benjamin, archt., 99
Bridges, archt., 83
Brighton—
  Norfolk Square, 133
Bristol—
  Berkeley Square, 85
  Castle, The, 86
  Churches:
    St. Mary-le-Port, 82
    St. Nicholas', 83
    St. Peter's, 82
    Temple Church, 83
  (*See also* Clifton)
  Dutch House, 84
  Freemasons' Hall, 86
  Merchant Venturers' Almshouses, 85
  Museum, 87
  Park Street, 86
  St. Peter's Hospital, 84
  Upper Arcade, 87
Bromley, Kent—
  Parish Church, 134
Brooker, Joshua, archt., 112
Brooks, James, archt., 41, 42
Burlington, Lord, 55
Burne-Jones, Sir Edward, 118
Burton, Decimus, archt., 16
Butterfield, William, archt., 46, 47

Caine, Captain, builder, 7
Canterbury—
  Burgate Street, 129
  The Deanery, 129
Cardiff—
  Llandaff Cathedral, 138
Chambers, Sir William, archt., 63
Chetham, Humphrey, 108, 110
Clarke, J., archt., 8
Clayton, A. B., archt., 44
Clifton—
  All Saints' Church, 89
  Holy Trinity Church, Hotwells, 90
  Parish Church, 88-89
Cockerell, S. P., archt., 32, 73, 86, 90
Comper, J. N., archt., 42, 46
Cooper, Sir Edwin, archt., 59
Coventry—
  Churches:
    Christ Church, 93
    St. Michael's (the Cathedral), 92-93
  Cook Street, house in, 96
  Cox Street, house in, 96
  Ford's Hospital, 94
  Greyfriars Lane, house in, 96
  Little Park Street, house in, 96
  Priory Row, 95
Croydon—
  St. Anselm's School, 125
Cubitt, builder, 7

Dance, George (the younger), archt., 6
Deptford (*see* London)
Dibden, Hants—
  Parish Church, 136
Donaldson, T. L., archt., 78

Dover—
  Round House, 132
  St. James-the-less, Church of, 132
Dowbiggin, Lancelot, archt., 41

East Ham—
  Parish Church, 126
Eden, F. C., archt., 89
Edge, Charles, archt., 119
Elmes, James, archt., 8
Eton College, 128-129
Eveleigh, archt., 105

Fletcher, Sir Banister, archt., 20
Foster, James, archt., 87, 88
  John, archt., 113, 114, 115, 116
Foster and Wood, archts., 87
Foulston, John, archt., 102, 106

Gandy-Deering, archt., 78
Garner, T., archt., 20
Gibbons, Grinling, 17, 29
Gibbs, James, archt., 31, 37
Godwin, E. W., archt., 104
Goodwin, Francis, archt., 98
Gough, A. R., archt., 90
Great Coggeshall, Essex—
  Parish Church, 135
Greenwich (*see* London)

Hakewill, Henry, archt., 60
Hamp, Stanley, archt., 75
Hardwick, Philip, archt., 41, 77
  Thomas, archt., 41
Harrison, Thomas, archt., 112
Hawksmore, Nicholas, archt., 33, 34, 65
Hill, William, archt., 100
Horne, Herbert, archt., 50
Hull—
  Albion Street Congregational Chapel, 124
  Charterhouse, 123
  George Street, 124
  Master Mariners' Almshouses, 123
  Pilot Office, 124
  Warehouse (Museum), 125
Hutchinson, archt., 93, 119

I'Anson, Edward, archt., 10
Isaacs and Florence, archts., 59

James, John, archt., 33, 38
Jones, Sir Horace, archt., 6
Jones, Inigo, archt., 7
Juxon, Archbishop, 11

Kay, Joseph, archt., 68
Kent, William, archt., 52, 64
Knightly, T. E., archt., 79, 80

Liverpool—
  Blue Coat Chambers, 115
  Burlington House, 111
  Churches:
    St. Catherine's, 113
    St. Luke's, 114
    St. Lambeth Palace, 113
    St. Nicholas', 112
  Custom House, 115
  Goree Warehouses, 116
  St. George's Crescent, 116
Llandaff Cathedral, 138
Lockwood and Mawson, archts., 51, 124
London—
  Adelphi, John Street, 75
  Albany Street, 75
  Arts Club, The, 66
  Bakers' Hall, 8
  Barbers' Hall, 7
  Brewers' Hall, 7
  Bridgewater House, 73
  Buckingham Palace, 63
  Burlington Arcade, 76
  Carlton House Terrace, 71
  Charterhouse, 56
  Chelsea Hospital, 65
  Churches:
    All Hallows, Barking, 15
    All Souls, Langham Place, 43

London—Churches (continued)—
  Ascension, Chapel of the, 50
  Austin Friars (Dutch Church), 14
  Chelsea Old Church, 14
  Christchurch, Albany Street, 75
  Christchurch, Newgate Street, 26
  Holy Trinity, Minories, 41
  Our Lady of Victories, Kensington, 50
  St. Alban's, Holborn, 46
  St. Alban's, Wood Street, 28
  St. Alphege's, Greenwich, 33
  St. Andrew's, Holborn, 20
  St. Andrew's-by-the-Wardrobe, 20
  St. Anne and St. Agnes, Gresham Street, 20
  St. Anne's, Soho, 32
  St. Augustine's, Watling Street, 30
  St. Barnabas', Finsbury, 41
  St. Bride's, Fleet Street, 22-23
  St. Clement's, City Road, 47
  St. Clement Danes, 31
  St. George's Cathedral, Lambeth, 45
  St. George's-in-the-East, 34-35
  St. Giles's, Cripplegate, 15
  St. James's, Piccadilly, 28-29
  St. James's, West Hackney, 44
  St. John-the-Divine, Kennington, 48
  St. John's, Horselydown, 38-39
  St. John's, Red Lion Square, 49
  St. John's, Smith Square, 36
  St. John's, Waterloo Road, 42
  St. Lawrence, Jewry, 17
  St. Mark's, Kennington, 44
  St. Martin's-in-the-Fields, 37
  St. Mary Abchurch, 24
  St. Mary's, Aldermanbury, 21
  St. Mary-le-Bow, 18-19
  St. Mary's, Haggerston, 42
  St. Mary's, Islington, 40-41
  St. Mildred's, Bread Street, 30
  St. Nicholas, Cole Abbey, 27
  St. Nicholas', Deptford, 33
  St. Olave's, Hart Street, 15
  St. Stephen's, Coleman Street, 17
  St. Stephen's, Walbrook, 24-25
  St. Swithun's, Cannon Street, 24
  St. Vedast's, Foster Lane, 26
  Temple Church, 16
  City Temple, 50-51
  Commons, House of, 52-54
  Crown Office Row, Temple, 59
  Deanery, The, Westminster, 54-55
  Essex Street arch, 74
  Featherstone Buildings, Holborn, 67
  Finsbury Square, 68
  Fishmongers' Hall, 9
  Garland's Hotel, Suffolk Street, 70
  Girdlers' Hall, 8-9
  Gray's Inn, 58-59
  Guilford Street, Bloomsbury, 73
  Guildhall, 6
  Haberdashers' Hall, 7
  Harpur Street, Theobald's Road, 67
  Holland House, 64
  Kensington Palace, 64
  Lambeth Palace, 11
  Lincoln's Inn Fields, No. 35, 72
  Lincoln's Inn, Stone Buildings, 57
  Mecklenburgh Square, 68
  Merchant Taylors' Hall, 10
  Metropolitan Tabernacle, Newington Butts, 50
  Middle Temple Hall, 60-61
  Middle Temple Lane, 59
  Montagu House (Portman House), 69
  Naval and Military Club, 66
  Pagani's Restaurant, 79
  Parish Clerks' Hall, 9
  Park Crescent, Regents Park, 70
  Park Village, 72
  Parliament, Houses of, 52-53
  Pepys monument, 15
  Portman House, 69
  Pump Court, Temple, 59
  Queen's Hall, The, 80
  Ring, The, Blackfriars Road, 76
  St. Katherine's Dock Office, 77
  St. Paul's Cathedral, 12-13

London (continued)—
  St. Paul's Chapter House, 74
  Serjeant's Inn, 57
  Sloane, Sir Hans, monument to, 14
  Somerset House, 63
  Stationers' Hall, 8-9
  Stone Buildings, Lincoln's Inn, 57
  Suffolk Street, Haymarket, 70
  Temple, The, 59
  Times Building, The, 78
  Tower, The, 56
  Tower Hill, warehouse on, 77
  Trinity House, 62
  University College, Gower Street, 78
  Westminster Abbey, 54-55
    Hall, 52-53
    School, 55

Manchester—
  Chetham's Hospital, 110
  Collegiate Church (the Cathedral), 108
  Cross Street Unitarian Chapel, 110
  Free Trade Hall, 109
  Royal Exchange, 109
  Victoria Buildings and Hotel, 108
  Wellington Inn, 107
Marshall, Joshua, mason, 17, 24
Martineau, E. H., archt., 7
Mee and Webb, archts., 120
Mills and Murgatroyd, archts., 109
Mitchell, John, joiner, 82
Moore, Temple, archt., 29
Mountain, Richard, archt., 123
Mylne, W. C., archt., 9

Nash, John, archt., 37, 42, 43, 69, 70-72, 75, 132
Newmarket—
  High Street and " White Hart," 131
Norman and Hine, archts., 104
Nottingham—
  St. John's Church, Lean Side, 122
  University College, 122

Pakefield, Suffolk—
  Parish Church, 136
Peacock, James, archt., 68
Pearson, J. L., archt., 49

Pennethorne, James, archt., 72, 75
Phillips, carpenter, 9
Phipps, C. J., archt., 80
Pite, Beresford, archt., 79
Plymouth—
  Athenaeum, 106
  Chapels :
    George Street Baptist, 104
    Norley Chapel, 103
    Treville Street Unitarian, 103
  Churches :
    Charles Church, 103
    St. Andrew's, 102
  Guildhall, 104-105
  Lockyer Street, 106
  Old Guildhall, 105
  Princess Square, 106
  Royal Hotel, 106
Portsea (*see* Portsmouth).
Portsmouth—
  Churches :
    Garrison Church, 98
    St. George's, Portsea, 98
    St. Paul's, Southsea, 97-98
  Guildhall, 100
  High Street, 99
  High Street Unitarian Chapel, 100
  Museum, 99
  St. Paul's Square, Southsea, 99
Prichard, John, archt., 138
Pugin, A. W. N., archt., 45, 52, 54

Reveley, Willey, archt., 120
Rickman, Thomas, archt., 93, 119
Roberts, Henry, archt., 9
Robinson, Thomas, smith, 74
Roper, D. R., archt., 44
Rossi, sculptor, 12

St. Aubyn, archt., 59
Sanderstead, Surrey—
  All Saints' Church, 134
Savage, James, archt., 16, 60
Scott, Sir George Gilbert, archt., 9, 28, 102, 122
Scott, Sir Giles, archt., 144
Searle and Hayes, archts., 50
Shields, Frederic, painter, 50

Shoppee, C. J., archt., 8
Smirke, Sir Robert, archt., 16, 44, 59
Smirke, Sydney, archt., 16, 59
Southsea (*see* Portsmouth)
Southampton—
  All Saints' Church, 120
  Holy Rood Church, 120
  St. Mary's Church, 121
Southwick, Sussex—
  Parish Church, 137
Stevens, Alfred, painter, 12
Stow-Bedon, Norfolk—
  Parish Church, 136
Street, George Edmund, archt., 48, 89, 121
Stuart, James, archt., 69
Sturry, Kent, 133
Swansea—
  Market, 126
  Wesley Chapel, 126

Taylor, Sir Robert, archt., 57, 72
Telford, Thomas, 77
Teulon, S. S., archt., 20
Theed, William, sculptor, 108
Thornhill, Sir James, painter, 24
Thorpe, John, archt., 64
Timbrell, carpenter, 59

Vardy, John, archt., 66

Walter, John, of *The Times*, 78
Walters, Edward, archt., 42, 109
Ware, Samuel, archt., 76
Watts, G. F., painter, 12
Wigg, Francis, archt., 59
Wilcox, carpenter, 29
Wilkins, William, archt., 78
Wilson, J. B. and Moxham, G., archts, 126
Workman and Lowe, builders, 9
Worley, Charles H., archt., 79
Wren, Sir Christopher, archt., 6, 7, 12, 17, 18, 20, 21, 23, 24, 25, 26, 27, 28, 29, 30, 31, 53, 64, 65, 74,
Wyatt, Samuel, archt., 52, 62

Yarmouth—
  Greyfriars, 130
  Tollhouse, 130

www.ingramcontent.com/pod-product-compliance
Lightning Source LLC
Chambersburg PA
CBHW021759230426
43669CB00006B/136